3D Biometrics

David Zhang · Guangming Lu

3D Biometrics

Systems and Applications

Springer

David Zhang
Biometrics Research Centre
The Hong Kong Polytechnic University
Hung Hom
Hong Kong

Guangming Lu
Shenzhen Graduate School
Harbin Institute of Technology
Shenzhen
China

ISBN 978-1-4899-9342-7   ISBN 978-1-4614-7400-5   (eBook)
DOI 10.1007/978-1-4614-7400-5
Springer New York Heidelberg Dordrecht London

Printed on acid-free paper

Springer is part of Springer Science+Business Media (www.springer.com)

Preface

Recently, biometric technology has been one of the hottest research topics in the IT field, because of the demands for accurate personal identification or verification to solve security problems in various applications, such as, e-commence, Internet banking, access control, immigration, law enforcement and so on. Especially after the 9/11 terrorist attacks, the interest in biometrics-based security solutions and applications has increased dramatically.

Although a lot of traditional biometric technologies and systems such as fingerprint, face, palmprint, voice and signature have been greatly developed over the past decades, they are application dependent and still have some limitations. 3D biometric technologies are emerging for high security requirement with their advantages: 3D biometrics are much more robust to illumination and pose variations from 2D biometrics; 3D range data may offer a richer information source for feature extraction. Besides, It can fuse with 2D biometrics to enhance the system accuracy; 3D biometric systems are more robust to attack, since 3D information is more difficult to be duplicated or counterfeited.

With the development of 3D imaging techniques, it is possible to capture real-time 3D biometric characteristics. Recently, 3D techniques have been used in biometric authentication, such as 3D face, 3D fingerprint, 3D palmprint and 3D ear recognition, and some commercial 3D biometric systems have been pushed into the market already.

Our team certainly regards 3D biometrics as a very potential research field, and has worked on it since 2005. We are the first group that developed the 3D palmprint technology and system, and our first technical paper of 3D palmprint, "Three dimensional palmprint recognition using structured light imaging", was published in 2008. We built the first 3D palmprint database (PolyU 3D Palmprint Database), which contains 8,000 samples collected from 400 different palms, and have published it online since 2010. Until now this database has been downloaded by many researchers. This work was followed by more extensive investigations into 3D palmprint technology, and this research has now evolved to other 3D biometric fields, such as 3D ear by line structured light, 3D fingerprint by multi-view imaging and 3D face by time-of-flight methods. Then, a number of algorithms have been proposed for these 3D biometric technologies, including calibration, 3D modeling, segmentation approaches, feature extraction methodologies, matching strategies and classification ideas. Both explosion of interest and diversity of approaches have been reflected in the wide range of recently published technical papers.

This book seeks to gather and present current knowledge relevant to the basic concepts, definition and features of 3D biometric technology in a unified way, and demonstrates some 3D biometric identification system prototypes. We hope thereby to provide readers with a concrete survey of the field in one volume. Selected chapters provide in-depth guides to specific 3D imaging methods, algorithm designs and implementations.

This book provides a comprehensive introduction to 3D biometric technologies. It is suitable for different levels of readers: those who want to learn more about 3D biometric technology, and those who wish to understand, participate in and/or develop a 3D biometric authentication system. We have tried to keep explanations elementary without sacrificing depth of coverage or mathematical rigor. Part I of this book explains the background of 3D biometrics. 3D ear recognition by line structured light is introduced in Part II. Part III presents 3D palmprint technologies by using modulated structured light imaging. 3D fingerprint identification by multi-view imaging and 3D face verification by time-of-flight method are developed in Part IV and Part V, respectively.

This book is a comprehensive introduction to both theoretical issues and practical implementation in 3D biometric authentication. It will serve as a textbook or as a useful reference for graduate students and researchers in the fields of computer science, electrical engineering, systems science and information technology. Researchers and practitioners in industry and R&D laboratories working on security system design, biometrics, immigration, law enforcement, control and pattern recognition will also find much of interest in this book.

The work is supported by the NSFC funds under project Nos. 61272292, 61271344 and 61020106004, Shenzhen Fundamental Research fund JC201005260184A, and Key Laboratory of Network Oriented Intelligent Computation, Shenzhen, China.

Hong Kong, July 2012   David Zhang
China   Guangming Lu

Contents

Part II   3D Ear Recognition Based on Line Structured Light

Part I
Background of 3D Biometrics

Chapter 1
Overview

Abstract Recently, biometrics technology is one of the hot research topics in the IT field because of the demands for accurate personal identification or verification to solve security problems in various applications. This chapter gives an all-around introduction to biometrics technologies, and the new trend: 3D biometrics.

Keywords Biometrics • 3D biometrics • Identification • Verification

1.1 The Need for Biometrics

Biometrics lies in the heart of today's society. There has been an ever-growing need to automatically authenticate individuals at various occasions in our modern and automated society, such as information confidentiality, homeland security, and computer security. Traditional knowledge based or token based personal identification or verification is so unreliable, inconvenient, and inefficient, which is incapable to meet such a fast-pacing society. Knowledge-based approaches use "something that you know" to make a personal identification, such as password and personal identity number. Token-based approaches use "something that you have" to make a personal identification, such as passport or ID card. Since those approaches are not based on any inherent attributes of an individual to make the identification, it is unable to differentiate between an authorized person and an impostor who fraudulently acquires the "token" or "knowledge" of the authorized person. This is why biometrics identification or verification system started to be more focused in the recent years.

Biometrics involves identifying an individual based on his/her physiological or behavioral characteristics. Many parts of our body and various behaviors are embedded such information for personal identification. In fact, using biometrics for person authentication is not new, which has been implemented over

D. Zhang and G. Lu, *3D Biometrics*, DOI: 10.1007/978-1-4614-7400-5_1,

1,000 years, numerous research efforts have been put on this subject resulting in developing various techniques related to signal acquisition, feature extraction, matching and classification. Most importantly, various biometrics systems including, fingerprint, iris, hand geometry, voice and face recognition systems have been deployed for various applications (Jain et al. 1999).

According to the annual report of the International Biometric Group (IBG, New York), the market for biometrics technologies increased around 25 % each year in the past 3 years. Figure 1.1a shows predicted total revenues of biometrics for

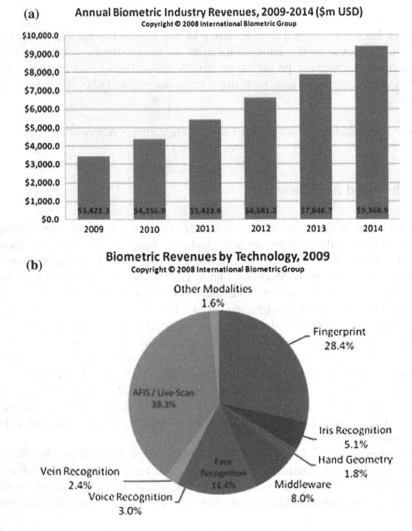

Fig. 1.1 **a** Total biometrics revenues prediction in 2009–2014. **b** Comparative market sharing by biometrics technologies in 2009 (International Biometric Group 2009)

2009–2014. Figure 1.1b shows Comparative Market Share by different biometrics technologies for the year 2009.

1.1.1 Biometrics System Architecture

A biometrics system has four major components: User Interface Module, Acquisition Module, Recognition Module and External Module. Figure 1.2 shows the breakdown of each module of the biometrics authentication system. The functions of each component are listed below:

1. User Interface Module provides an interface between the system and users for the smooth authentication operation. It is crucial to develop a good user interface such that users are happy to use the device.
2. Acquisition Module is the channel for the biometrics traits to be acquired for the further processing.
3. Recognition Module is the key part of our system, which will determine whether a user is authenticated. It consists of image pre-processing, feature extraction, template creation, database updating, and matching. Then it gives an identification/verification result.
4. External Module receives the signal come from the recognition module, to allow some operations to be performed, or denied the operations requested. This module actually is an interfacing component, which may be connected to another hardware components or software components.

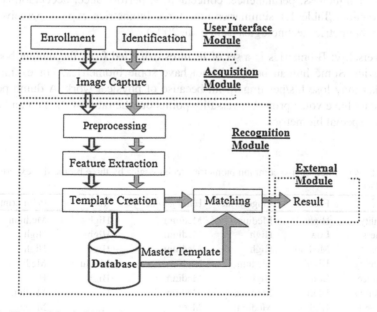

Fig. 1.2 The modules in biometrics system

1.1.2 Operation Mode of a Biometrics System

A biometrics system is usually operated in three modes: enrollment, identification and verification. But some systems only have either identification or verification modes.

Enrollment—Before a user can be verified or identified by the system, he/she must be enrolled by the biometrics system. The user's biometrics data is captured, preprocessed and feature extracted as shown in stages 1–3 of Fig. 1.2. Then, the user's template is stored in a database or file system.

Identification—This refers to the identification of a user based solely on his/her biometrics information, without any prior knowledge about the identity of a user. Sometimes it is referred to 1-to-many matching, or recognition. It will go through stages 1–3 to create an identification template. Then the system will retrieve all the templates from the database for the feature matching. A result of success or failure is given finally. Generally, accuracy decreases as the size of the database grows.

Verification—This requires that an identity (ID card, smart card or ID number) is claimed, and then a matching of the verification template with master templates is performed to verify the person's identity claim. Sometimes verification is referred to a 1-to-1 matching, or authentication.

1.1.3 Evaluation of Biometrics and Biometrics System

Seven factors affect the determination of a biometrics identifier, including: universality, uniqueness, permanence, collectability, performance, acceptability, and circumvention. Table 1.1 summaries how three biometrics experts perceive five common biometrics technologies (Jain et al. 2004).

1. Universality: Biometrics is a set of features extracted from the human body or behavior. Some human beings do not have some biometrics. For example, a worker may lose his/her fingerprint because of physical work. A dumb person does not have voice print. Universality points out the ratio of the human beings with a special biometrics.

Table 1.1 Perception of five common biometrics technologies by three biometrics experts (Jain et al. 2004)

	Face	Fingerprint	Hand geometry	Iris	Palm print
Universality	High	Medium	Medium	High	Medium
Uniqueness	Low	High	Medium	High	High
Permanence	Medium	High	Medium	High	High
Collectability	High	Medium	High	Medium	Medium
Performance	Low	High	Medium	High	High
Acceptability	High	Medium	Medium	Low	Medium
Circumvention	High	Medium	Medium	High	Medium

2. Uniqueness: If a biometrics is unique, it can be used to completely distinguish any two persons in the world. The identical twins with the same genetic genotype are one of the important test for uniqueness. Observing the similarity of a biometrics in a large database is also an important indicator for uniqueness.

3. Permanence: Many biometrics will change time by time, such as voice print, face. Iris and fingerprint, which are stable in a long period of time, are relative permanence. Permanence is described by the stability of a biometrics.

4. Collectability: Although some biometrics have high permanence, uniqueness and universality, it cannot be used for public because of collectability. If the data collection process is too complex or requires high cost input devices, the collectability of this biometrics is low. DNA and retina suffer from this problem.

5. Performance: The term "Performance" is referred to accuracy, which is defined by two terms, (1) False Acceptance Rate (FAR) and (2) False Rejection Rate (FRR) which are controlled by a threshold. Reducing FAR (FRR) has to increase FRR (FAR). Equal Error Rate (EER) or crossover rate also refers accuracy.

6. Acceptability: To be a computer scientist, we should try our best to produce a user-friendly biometrics system. In fact, almost all the current biometrics systems are not physically intrusive to users but, some of them such as, retina-based recognition system, are psychologically invasive system. Retina-based recognition system requires a user to put his/her eye very close to the equipment and then infrared light passes through his/her eye in order to illuminate his/her retina for capturing an image (Miller 1994; Zhang 2000a, b; Mexican Government 2003.

7. Circumvention: The term "Circumvention" refers to how easy it is to fool the system by using an artifact or substitute.

1.2 Different Biometrics Technologies

At present, the biometrics technologies can be grouped into three categories based on the signal dimension: 1D, 2D, and 3D. 1D biometrics mainly contains voice and signature recognition technologies. 2D biometrics technologies are widely used, such as iris, face, fingerprint, hand geometry, palm print, and other biometrics technologies based on 2D images. Recently, researchers have focused on the use of three dimensional data as a source of distinguishing features for personal identification, such as 3D face recognition, 3D palm print recognition, and 3D fingerprint (Woodard 2006).

Each existing system has its own strengths and limitations. There is no perfect biometrics system until now, and the question of which one is better depends on the application. The following shows different types of biometrics technologies and systems available on the market.

1.2.1 One-Dimension Technologies

1.2.1.1 Voice Recognition Technology

Voice (speaker) recognition consists of identification and verification of the voice (Barbu 2009). Voice recognition methods encompass text-dependent and text-independent methods (Cole et al. 1997; Sammut and Squires 1995). The text-dependent methods discriminate the voice by the same utterance, such as specifically determined words, numbers or phrases. The text-independent methods, on the other hand, recognize the voice no matter what form of words or numbers the speakers' provide. There will be waveform formed by the voice when a person speaks a word or number. The waveform is known as a voice pattern which like the fingerprints or other physical features is unique. Although everyone has a different pitch, which can be considered physical features, the human voice is classified into behavioral biometrics identifier (Barbu 2009; Rashid 2008). However, it is not only the qualities of the microphone and the communication channel but also the aging, medical conditions or even emotional status can affect the behavioral part of the speech of a person (Mastali 2010).

Some researchers have developed the smart home which based on the voice recognition (Baygin 2012). In the home, you can control the equipment by your voice such as turning on a lamp, close the curtain and so on. The voice recognition is also used in the e-commerce to solve the problems facing the mobile e-commerce (Yang 2011).

1.2.1.2 Signature Recognition Technology

Signature recognition represents an important biometrics recognition field. It has a long tradition in many common commercial fields. The technology considered to be a behavioral biometrics generally be divided into two types, namely on-line and off-line signature recognition. The on-line signature recognition is mainly based on the 1-Dimensional features such as pressure, velocity, acceleration of the signature, etc. But the off-line signature recognition is mainly based on the static images of the signed words. Some on-line signature recognitions not only deal with the time domain information described above but also process the static signature images.

The online signature verification system developed by Prof. Berrin Yanikoglu and Alisher Kholmatov has won the first place at The First International Signature Verification Competition (SVC 2004) organized in conjunction with the First International Conference on Biometric Authentication (ICBA 2004). The recognition rates were 2.8 % equal error rate for skilled forgery tests (Signature Verification 2002).

Li and Zhang proposed a method for stroke-based online signature verification using null component analysis (NCA) and principal component analysis (PCA). After the segmentation and flexible matching of the signature, they extracted stable segments from each reference signature in order that the segment sequences had the same length. The reference set of feature vectors are transformed and separated into null components (NCs) and principal components (PCs) by K-L transform. Experiments on a data set containing a total 1,410 signatures of 94 signers showed that the NCA/PCA-based online signature verification method could achieve better results. The best result yielded an equal error rate of 1.9 %. Figure 1.3 shows an original signature in 2-D and 3-D with time information, and Fig. 1.4 shows an optimal matching path between two genuine signatures (Li et al. 2006).

1.2.2 Two-Dimension Technologies

1.2.2.1 Iris Recognition System

Iris recognition is one of the most effective biometrics technologies, being able to accurately identify the identities of more than thousand persons in real-time

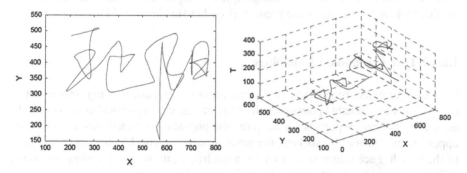

Fig. 1.3 An original signature is shown in 2-D and 3-D with time information (Li et al. 2006)

Fig. 1.4 An optimal matching path between two genuine signatures (Li et al. 2006)

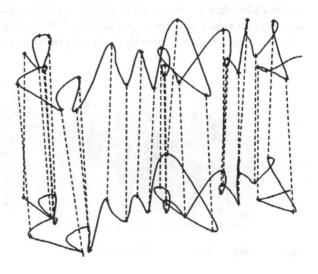

(Pankanti et al. 2000; Zhang 2000a, b; Jain and Pankanti 2001; Jain et al. 2004; Daugman 1993; Daugman 2003). The iris is the ring that surrounds the pupil. A camera using visible and infrared light scans the iris and creates a feature template based on characteristics of the iris tissue, such as rings, furrows, and freckles.

There are some public iris databases contributed by some research teams to promote the research, such as Iris images database of CASIA V1.0 (CASIA-IrisV1), which is presented by National Laboratory of Pattern Recognition (NLPR), Institute of Automation (IA), Chinese Academy of Sciences (CAS). Figure 1.5 shows some typical images from this dataset.

An iris recognition system such as IrisAccess™ from Iridian Technologies, Inc. provides a very high level of accuracy and security (Iridian Technologies 2012). Its scalability and fast processing power fulfils the strict requirements of today's marketplace but it is expensive and users regard it as intrusive. It is suitable for high security areas such as nuclear plants or airplane control rooms. On the other hand, it is not appropriate in areas which require frequent authentication processes, such as logging onto a computer. Another iris recognition system such as InSight™ VM from AOptix Technologies, Inc. (AOptix 2010) has been applied in the Gatwick airport to ensure the security (London Heathrow 2011).

1.2.2.2 Face Recognition Technology

Compared to others biometrics, face verification is low cost, needing only a camera mounted in a suitable position such as the entrance of a physical access control area. For verification purposes it captures the physical characteristics such as the upper outlines of the eye sockets, the areas around the cheekbones, and the sides of the mouth. Face-scanning is suitable in environments where screening and surveillance are required with minimal interference with passengers.

The State of Virginia in the U.S. has installed face-recognition cameras on Virginia's beaches to automatically record and compare their faces with images of suspected criminals and runaways (McGuire 2003). However, the user acceptance of facial scanning is lower than that of fingerprints, according to a IBG Report.

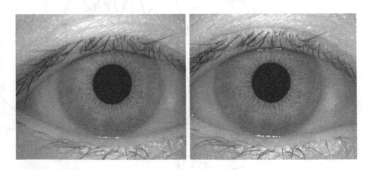

Fig. 1.5 Example iris images in CASIA-IrisV1

Many airports have installed facial recognition system to heighten their security. Following is a successful trial alongside the UK Border Agency. London Heathrow Airport will implement infrared facial recognition called the Aurora Imaging Recognition system, which can confirm the identity of the average passenger in five seconds to heighten security (London Heathrow 2011). All passengers will be checked when they pass the Terminals and in the same time present their boarding pass before they boarding their flight. This scheme ensures that a traveler cannot swap tickets with others.

1.2.2.3 Fingerprint Recognition Technology

Automatic fingerprint identification began in the early 1970s. At that time, fingerprint verification had been used for law enforcement. From 1980s, the rapid development of personal computer and fingerprint scanner; consequently, fingerprint identification started to be used for non-criminal applications (Zhang 2000a, b). Current fingerprint systems utilize minutiae and singular points as the features (Jain and Pankanti 2001; Jain et al. 2004; Daugman 1993, 2003; McGuire 2002).

The most promising minutiae points are extracted from an image to create a template, usually between 250 to 1,000 bytes in size (Jain et al. 1997; Jain and Pankanti 2001; Ratha et al. 1996; Karu and Jain 1996; Cappelli et al. 1999; Maio and Maltoni 1997; Berry 1994). It is the most widely used biometrics technology in the world. Its small chip size, ease of acquisition and high accuracy make it the most popular biometrics technology since the 1980s. However, some people may have fingerprints worn away due to hand work and some old people may have many small creases around their fingerprints, lowering the system's performance. In addition, the fingerprint acquisition process is sometimes associated with criminality, causing some users to feel uncomfortable with it.

1.2.2.4 Palm Print Recognition Technology

Palm print is concerned with the inner surface of a hand and looks at line patterns and surface shape. A palm is covered with the same kind of skin as the fingertips and it is larger than a fingertip in size. Therefore, it is quite natural to think of using palm print to recognize a person. Because of the rich features including texture, principal lines and wrinkles on palm prints, it is believed that they contain enough stable and distinctive information for separating an individual from a large population.

There have been some companies, including NEC and PRINTRAK, which have developed several palm print systems for criminal applications (Jain et al. 1997; Miller 1994). On the basis of fingerprint technology, their systems exploit high resolution palm print images to extract the detailed features like minutiae for matching the latent prints. Such approach is not suitable for developing a palm print authentication system for civil applications, which requires a fast, accurate and reliable method for the personal identification. The Hong Kong Polytechnic University developed a novel palm print authentication system to fulfill such requirements, as shown in Fig. 1.6. Figure 1.6a shows a CCD camera-based 2D palm print acquisition device, and Fig. 1.6b is a palm print image collected by this device (Jain et al. 1997).

1.2.2.5 Hand Geometry Recognition Technology

Hand geometry recognition is one of the oldest biometrics technologies used for automatic person authentication. Hand geometry requires only small feature size, including the length, width, thickness and surface area of the hand or fingers of a user (Sanchez-Reillo et al. 2000; Kumar and Zhang 2006), as shown in Fig. 1.7.

There is a project called INSPASS (Immigration and Naturalization Service Passenger Accelerated Service System) which allows frequent travelers to use 3D hand geometry at several international airports such as Los Angeles, Washington, and New York. Qualified passengers enroll in the service to receive a magnetic stripe card with their hand features encoded. Then they can simply swipe their card, place their hand on the interface panel, and proceed to the customs gate to avoid the long airport queues. Several housing construction companies in Hong Kong have adopted the Hand Geometry for the employee attendance record in their construction sites. A smart card is used to store the hand shape information and employee details. Employees verify their identities by their hand features against the features stored in the smart card as they enter or exit the construc- tion site. This measure supports control of access to sites and aids in wage calculations.

Hand geometry has several advantages over other biometrics, including small feature size, less invasive, more convenient and low cost of computation as a result of using low-resolution images (Sanchez-Reillo et al. 2000; Sanchez-Reillo and Sanchez-Marcos 2000). But the current hand geometry system suffers from high

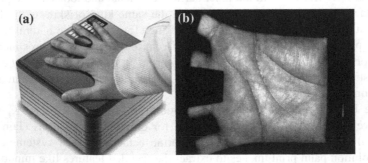

Fig. 1.6 **a** CCD camera-based 2D palm print acquistion device. **b** A palm print image (*right*) that is collected by the device

Fig. 1.7 Hand-shape features extracted from the hand images

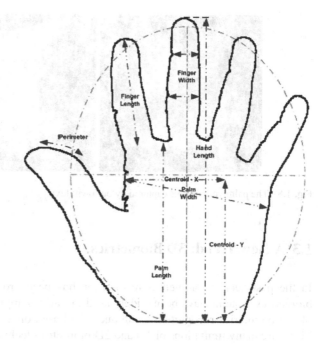

cost and low accuracy (Pankanti et al. 2000). In addition, uniqueness of the hand features is not guaranteed, making it unfavorable to be used in one-to-many identification applications.

1.2.2.6 Palm Vein Patterns

Fujitsu Laboratories Limited developed a new type of biometrics authentication system (PalmSecure™) which verifies a person's identity by the pattern of the veins in his/her palms (Fujitsu Laboratories Limited 2002). The PalmSecure™ works by capturing a person's vein pattern image while radiating it with near-infrared rays. The scanning process is extremely fast and does not involve any contact between the sensor and the person being scanned (Fujitsu Laboratories Limited 2011a, b, c).

Fujitsu piloted the palm vein recognition based solution from January to June of 2011 at Boca Ciego High School. The result was tremendous: a 98 % first-scan transaction success rate with wait times in their busy lunch lines dropping from 15 min to a paltry 7 min. Starting in the Fall of the 2011–2012 school year, all Pinellas County Schools will rely on the Fujitsu PalmSecure biometrics solution to handle cafeteria transaction for the tens of thousands of students, across all of their 46 middle and high-schools, who take part in their daily snack and lunch service program (Fujitsu Laboratories Limited 2011a, b, c). Figure 1.8 shows two palm vein images captured by our own device.

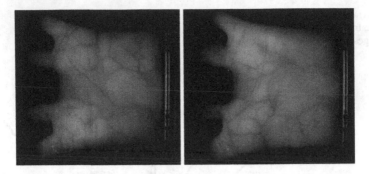

Fig. 1.8   The palm vein images captured by our own device

1.3 A New Trend: 3D Biometrics

In the past decade, biometrics recognition has been growing rapidly, and many biometrics systems have been widely used in various applications. However, most of the biometrics recognition techniques are based on 1D signals or 2D images. There are many limitations of 1D and 2D biometrics technologies until now:

- Fingerprints may be distorted and unreadable or unidentifiable if the person's fingertip has dirt on it, or if the finger is twisted during the process of finger-printing. In an ink fingerprint, twisting could cause the ink to blur, distorting the shape of the fingerprint and potentially making it unreadable.
- It is found that with age, the voice of a person differs. Also when the person has flu or throat infection the voice changes, or if there are too much noise in the environment this method may not authenticate correctly.
- For Iris recognition, if people affected with diabetes, the eyes get affected resulting in differences.
- The conventional 2D palm print recognition is a fast and effective personal authentication method, but 2D palm print images can be easily counterfeited.

Although 2D biometrics recognition techniques can achieve high accuracy, the 2D features can be easily counterfeited and much 3D feature structural information is lost. Therefore, it is of high interest to explore new biometrics recognition techniques: 3D Biometrics.

With the development of 3D techniques, it is possible to capture 3D characteristics in real time. Recently, 3D techniques have been used in biometrics authentication, such as 3D face, 3D fingerprint, 3D palmrpint and 3D ear recognition, and shown many advantages, such as:

- 3D biometrics is much more robust to illumination and pose variations than 2D biometrics.
- 3D range data may offer a richer information source for feature extraction. And usually it also can fuse with 2D biometrics to enhance the system accuracy.

- 3D biometrics systems are more robust to attack, since 3D information is more difficult to be duplicated or counterfeited.

3D biometrics technologies have been the new trend in this research field. There are some commercial devices which can obtain the 3D information of an object, such as Konica Minolta Vivid 9i/910 (Zhang 2000a, b), Cyberware whole body color 3D scanner (Mexican Government 2003), and so on. These commercial 3D scanners have high speed and accuracy, and can be used for 3D biometrics information collection.

1.4 Arrangement of this Book

In this book, we would like to summarize our 3D biometrics work. Its fifteen chapters are in five parts, covering 3D biometrics technology from the hardware design of 3D imaging systems, 3D modeling and reconstruction, 3D preprocessing algorithms, feature extraction, and matching.

PART I

This chapter introduces recent developments in biometrics technologies, some key concepts in biometrics, and the new trend: 3D biometrics technologies. Chapter 2 focuses on the main 3D imaging technologies and its application in 3D biometrics, and then we introduce some typical 3D ear, 3D palm print, 3D fingerprint, and 3D face data acquisition methods, 3D feature extraction and matching algorithms.

PART II

This part discusses 3D ear recognition by line structured light, including 3D ear acquisition system design, two significant characteristics in 3D ear, and 3D ear feature extraction and recognition methods.

PART III

This part has four chapters, focusing on the 3D palm print acquisition requirement and design, preprocessing methods, feature extraction and matching, and the classification of 3D palm prints. Chapter 6 introduces 3D palm print capturing system base on modulated structured light. Chapter 7 reports 3D information in palm print. Chapter 8 presents some methods for 3D palm print global feature extraction. And Chapter 9 proposes the joint line and orientation features in 3D palm print.

PART IV

This part contains three chapters. Chapter 10 introduces novel ideas for 3D fingerprint acquisition based on multi-view. Chapter 11 studies 3D fingerprint reconstruction technique from touchless multi-view fingerprint images. In Chap. 12, we propose a 3D fingerprint identification system.

PART V

This part includes two chapters. Chapter 13 focus on 3D face imaging by time-of-flight principle. In Chap. 14, we build a 3D face recognition system, develop and investigate suitable methods and techniques for 3D face recognition.

At the end of this book, a brief book review and future work are presented in Chap. 15.

References

AOptix (2010) AOptix announces new slim design iris recognition system. http://www.aoptix.com/news-events/press-releases/119-insight-vm-launch%20. Accessed 18 July 2012

Barbu T (2009) Comparing various voice recognition techniques. In: IEEE proceedings of the 5th conference on speech technology and human-computer dialogue, pp 1–6. doi: 10.1109/SPED.2009.5156172

Baygin M (2012) Real time voice recognition based smart home application. IEEE 20th signal processing and communications applications conference (SIU), pp 1–4. doi: 10.1109/SIU.2012.6204694

Berry J (1994) The history and development of fingerprinting. In: Lee HC, Gaensslen RE (eds) Advances in fingerprint technology. Florida, pp 1–39

Cappelli R, Lumini A, Maio D, Maltoni D (1999) Fingerprint classification by directional image partitioning. IEEE Trans Pattern Anal Mach Intell 21:402–421. doi:10.1109/34.765653

Cole RA, Mariani J, Uszkoret H, Zaenen A, Zue A (1997) Survey of the state of the art in human language technology. Cambridge University Press. ISBN: 0-521-59277-1

Daugman JG (1993) High confidence visual recognition of persons by a test of statistical independence. IEEE Trans Pattern Anal Mach Intell 15:1148–1161. doi:10.1109/34.244676

Daugman JG (2003) The importance of being random: statistical principles of iris recognition. Pattern Recogn 36(2):279–291

Fujitsu Laboratories Limited (2002) Biometric mouse with palm vein pattern recognition technology. http://pr.fujitsu.com/en/news/2002/08/28.html. Accessed 4 July, 2003

Fujitsu Laboratories Limited (2011a) Fujitsu PalmSecure selected as a "world changing idea" for 2011 by scientific American magazine. http://www.fujitsu.com/us/services/biometrics/palm-vein/SA_news.html. Accessed 18 July 2012

Fujitsu Laboratories Limited (2011b) PalmSecure—Fujitsu's world-leading authentication technology. http://www.fujitsu.com/emea/products/biometr-ics/intro.html. Accessed 18 July 2012

Fujitsu Laboratories Limited (2011c) What is PalmSecure™. http://www.fujitsu.com/global/services/solutions/biometrics/. Accessed 18 July 2012

International Biometric Group (2009) Biometrics market report. http://www.biometricgroup.com. Accessed 18 July 2012

Iridian Technologies, Inc. (2012) http://www.iriscan.com/. Accessed 20 July 2012

Jain A, Pankanti S (2001) Automated fingerprint identification and imaging systems. In: Lee HC, Gaensslen RE (eds) Advances in fingerprint technology, 2nd edn. New York. ISBN: 0849309239

Jain AK, Hong L, Bolle R (1997) On-line fingerprint verification. IEEE Trans Pattern Anal Mach Intell 19:302–314. doi:10.1109/34.587996

Jain AK, Bolle RM, Pankanti S (1999) Biometrics: personal identification in networked society. Boston Hardbound. ISBN: 978-0-7923-8345-1

Jain AK, Ross A, Prabhakar S (2004) An introduction to biometric recognition. IEEE Trans Circ Syst Video Technol, Spec Issue Image Video-Based Biometrics 14:4–20. doi:10.1109/TCSVT.2003.818349

Karu K, Jain AK (1996) Fingerprint classification. Pattern Recogn 29:389–404. doi:10.1016/0031-3203(95)00106-9

Kumar A, Zhang D (2006) Combining fingerprint, palmprint and hand-shape for user authentication. 18th international conference on pattern recognition, vol 4, pp 549–552. doi: 10.1109/ICPR.2006.383

Li B, Zhang D, Wang KQ (2006) Online signature verification based on null component analysis and principal component analysis. Pattern Anal Appl 8(4):345–356. doi:10.1007/s10044-005-0016-4

London Heathrow set to roll-out facial recognition security checks (2011) http://www.futuretravelexperience.com/2011/07/london-heathrow-set-to-roll-out-facial-recognition-security-checks/. Accessed 20 July 2012

Maio D, Maltoni D (1997) Direct gray-scale minutiae detection in fingerprints. IEEE Trans Pattern Anal Mach Intell 19:27–40. doi:10.1109/34.566808

Mastali N (2010) Authentication of subjects and devices using biometrics and identity management systems for persuasive mobile computing: a survey paper. IEEE 5th international conference on broadband and biomedical communications, pp 1–6. doi: 10.1109/IB2COM.2010.5723618

McGuire D (2002) Virginia beach installs face-recognition cameras. The Washington post. http://www.washingtonpost. com/ac2/wp-dyn/A19946-2002Jul3. Accessed 14 May 2003

Mexican Government (2003) Face recognition technology to eliminate duplicate voter registrations in upcoming presidential elections. http://www.shareholder.com/identix/ReleaseDetail.cfm?ReleaseID=53264. Accessed 15 May 2003

Miller B (1994) Vital signs of identity. IEEE Spectr 31:22–30. doi:10.1109/6.259484

Pankanti S, Bolle RM, Jain A (2000) Biometrics: the future of identification. IEEE Comput 33:46–49. doi:10.1109/2.820038

Rashid RA (2008) Security system using biometric technology: design and implementation of voice recognition system (VRS). IEEE Int Conf Comput Commun Eng 898–902. doi: 10.1109/ICCCE.2008.4580735

Ratha N, Karu K, Chen S, Jain AK (1996) A real-time matching system for large fingerprint databases. IEEE Trans Pattern Anal Mach Intell 18:799–813. doi:10.1109/34.531800

Sammut C, Squires B (1995) Automatic speaker recognition: an application of machine learning. In: Proceeding of the 12th international conference on machine learning. doi: 10.1.1.54.7372

Sanchez-Reillo R, Sanchez-Marcos A (2000) Access control system with hand geometry verification and smart cards. IEEE Aerosp Electron Syst Mag 15:45–48. doi:10.1109/62.825671

Sanchez-Reillo R, Sanchez-Avilla C, Gonzalez-Marcos A (2000) Biometric identification through hand geometry measurements. IEEE Trans Pattern Anal Mach Intell 22:1168–1171. doi:10.1109/34.879796

Signature Verification (2002) http://biometrics.sabanciuniv.edu/signatur-e.html. Accessed 18 July 2012

Woodard DL (2006) A comparison of 3D biometric modalities. Comput Vis Pattern Recogn Workshop 57–60. doi: 10.1109/CVPRW.2006.12

Yang W (2011) Application of voice recognition for mobile E-commerce security. IEEE 3rd Pacific-Asia conference on circuits, communications and system, pp 1–4. doi: 10.1109/PACCS.2011.5990286

Zhang D (2000a) Automated biometrics: technologies and systems. Hardbound, Boston. ISBN 0792378563

Zhang D (2000b) Biometrics technologies and applications. In: Proceedings of 1st international conference on image and graphics. Tianjin, China, pp 42–49

Chapter 2
3D Biometrics Technologies and Systems

Abstract Until now, many 3D scanning technologies have been used in 3D biometrics research field, such as structured light technique, time-of-flight method, and multi-view imaging. Each method has its own characteristics, and can be used for some specified 3D biometrics. This chapter gives an overall review of these 3D imaging technologies and their applications in biometrics.

Keywords Structure light • Multi-view imaging • Time-of-flight

2.1 Introduction

3D biometrics technologies have been greatly developed in the recent years with the support of 3D imaging technologies, as shown in Fig. 2.1 (Photon-X 2010), and now there are already some 3D biometrics systems on the market. These 3D biometrics systems are classified by the imaging techniques into several groups (as shown in Fig. 2.2). In the first level, it is classified by the number of cameras used, they are single view based and multi-view based technologies. The single view methods can be divided into two main categories: structured light and time-of-flight based 3D imaging technologies. And then, the structured light methods can be further divided into two types: laser and modulated light based 3D imaging technologies.

Each technology comes with its own limitations, advantages and costs. We should select the suitable technology for different biometrics object 3D image scanning, such as line structured light scanning is good for 3D ear acquisition, modulated structured light is used for 3D palmprint scanning, the face recognition using 3D camera and the 3D fingerprint acquisition system based on multi-view.

These systems are not only different in the imaging technologies, but also different in the feature extraction methods. In the following parts, the imaging technique and feature extraction will be presented separately.

D. Zhang and G. Lu, *3D Biometrics*, DOI: 10.1007/978-1-4614-7400-5_2,

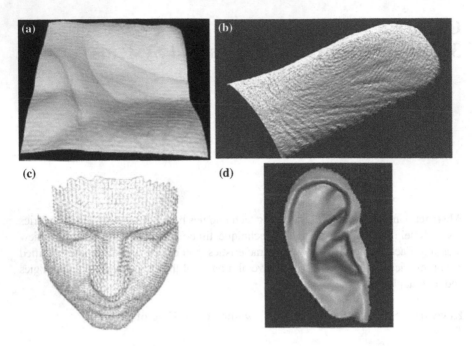

Fig. 2.1 Different 3D biometrics candidate characteristics. **a** 3D palmprint . **b** 3D fingerprint. **c** 3D face. **d** 3D ear

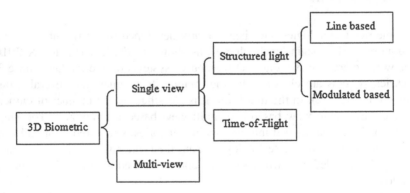

Fig. 2.2 The structure of the classification of the 3D biometrics technologies

2.2 Classification of 3D Biometrics Imaging Methods

Multi-view imaging is based on human visuals. It is high in speed and low in cost. However, it is hard to obtain high accuracy because it is difficult to find and match corresponding point pairs in two or more images. Structured-light imaging is widely used as a 3D imaging method for its high accuracy, speed and stability (Zhang 2004).

The structured light 3D vision measurement method was first seen in 1979s. Among many vision measurement methods, the structured light 3D vision has been widely used in the industry environments because of its characteristics, such as wide range, large field of view, higher precision, the grating image easy to get, strong real-time, active controlled and so on. Currently, the research on structured light 3D vision measurement method mainly focus on the projection model of the structured light, the processing algorithm and extraction of the sub-pixel grating center, the calibration method and so on.

2.2.1 Single View Imaging with Line Structured Light

The line structured light also called light band model. The light projected by a projector through a cylindrical mirror forms a narrow laser plane. When the laser plane contacts the object surface, there will be a grating on the object surface. The grating is modulated by the height of the object surface and the gap. It shows in the image is the grating become discontinues and distortional. The degree of the distortion is proportional to the height of the object surface and the discontinuity shows the physical gap on the object surface. Technique intuitively, the displacement showed by the grating information is proportional to the height of the object surface, the kink expresses the changing of the surface and the discontinuity shows the physical gap of the object surface. When the relative position of the projector and the camera is fixed, then the object surface 3D shape will be presented by the coordinate of the distortion 2D grating image. The task of the line structured light vision measurement is to get the 3D information of the object from the distortional grating image information.

Actually, the line structured light model is the extension of the point structured light model. The sight lines through the camera optical center intersect with the light projected by the projector and form many intersection points which show on the object surface are a lot of light spots on the grating. Thus it forms many triangular geometry constraint relationships like in the point structured light model. Obviously, the information in the line structured light model is more than that in the point structured light model. But its implementation complexity does not increase, so it has been widely used.

Multi-line structured light model is the extension of the line structured light model. The projector projects a few gratings. On one hand, it can handle more than one grating in one image, so it can improve the image processing efficiency. On the other hand, it realizes multi-grating cover on the object surface to increase the measurement information to get more range of the object surface height information. This model is so-called "grating structured model". Multi-grating can be formed by to be continued (Fig. 2.3).

2.2.2 Single View Imaging with Modulated Structured Light

The laser triangulation based 3D imaging technique has some shortcomings for the biometrics application. For example, the resolution of 3D cloud point may not

Fig. 2.3 Single view
imaging with line structured
light

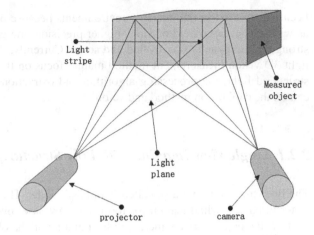

be high enough for the accuracy requirement in biometrics authentication. On the
other hand, if we improve the data resolution, the laser scanning speed must be
decreased and the requirement of real-time authentication is hard to meet.

With the above considerations and the requirements of accuracy and speed in
biometrics authentication, in the application of 3D recognition, we propose to use
modulated structured light imaging (Hung et al. 2000; Judge and Bryanston-Cross
1994) to establish the 3D model acquisition system.

Modulated light projector projects a continually changing light at the subject.
Usually the light source simply cycles its amplitude in a sinusoidal pattern. A camera detects the reflected light and the changes in the brightness of each pixel of the
image and the light phase to get the distance of the projected light. The structure is
shown in Fig. 2.4.

The use of modulated structured light in object surface measurement can be
traced back to more than two decades ago. Since then, it has been widely used

Fig. 2.4 Single view
imaging with modulated
structured light

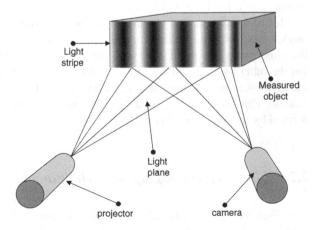

in many applications, such as 3-D object measurement, 3-D shape reconstruction, reverse engineering, etc. In structured light imaging, a light source projects some structured light patterns (stripes) onto the surface of the object. The reflected light is captured by a CCD camera and then a series of images is collected. After some calculation, the 3-D surface depth information of the object can be obtained.

In earlier times, parallel light such as laser or point light array were used. With the development of the light source techniques, liquid crystal light projectors have been successfully used as the light source.

Now, a cost-effective gray liquid crystal display (LCD) projector with LED light source is employed, and some shift light patterns are projected to the object surface. The modulated structured-light imaging is able to accurately measure the 3D surface of an object but use less time than line laser scanning. In the developed 3D palmprint system (Zhang et al. 2009), a LED projector will generate structured light stripes and project them to the palm when the user put his/her palm on the system,. In order to distinguish between stripes, they are coded with different brightness. The system uses a computer controlled LCD projector that can generate arbitrary stripe patterns. A series of grey level images of the palm with the stripes on it are captured by a CCD camera, and then the depth information of the palm surface is reconstructed from the stripe images. This device acquires 3D data in point cloud form. It also acquires a registered intensity image in the course of its normal operation.

2.2.3 Multi-View Imaging

Multi-view imaging has received a lot of interest recently. A number of camera arrays need to be built for multi-view imaging. For instance, Matusik et al. used 4 cameras for rendering using image-base visual hull (McMillan et al. 2000). Yang et al. had a 5-camera system for real-time rendering with the help of modern graphics hardware (Yang et al. 2002a); Schirmacher et al. built a 6-camera system for on-the-fly processing of generalized Lumigraphs (Schirmacher et al. 2001); Naemura et al. (Naemura et al. 2002) constructed a system of 16 cameras for real-time rendering. Several large arrays consisting of tens of cameras have also been built, such as the Stanford multi-camera array, the MIT distributed light field camera and the CMU 3D room. These three systems have 128, 64 and 49 cameras, respectively (Wilburn et al. 2002; Yang et al. 2002b; Kanade et al. 1998). In the above camera arrays, those with a small number of cameras can usually achieve real-time rendering. Large camera arrays, despite their increased viewpoint ranges, often have difficulty in achieving satisfactory rendering speed due to the large amount of data to be handled.

Chen et al. proposed a 3D fingerprint system based on multi-view. The use of multiple views enables the capture of the rolled-equivalent fingerprints that is faster than traditional rolling procedures, thereby increasing the usable fingerprint area. The different views can obtain by either different camera surrounding the

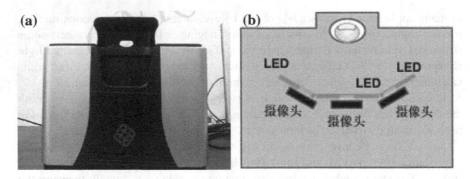

Fig. 2.5 Touchless multi-view fingerprint capture device and its schematic diagram

finger (Chen et al. 2006). The Hong Kong Polytechnic University also designed a multi-view fingerprint system as shown in Fig. 2.5a, and the schematic diagram of touchless multi-view fingerprint capture device is shown in Fig. 2.5b.

In Fig. 2.5, there are 5 cameras placed around the finger. From each acquired image, a silhouette is extracted. Knowing the position and orientation of each camera within a reference coordinate system, the 5 silhouettes are then projected into the 3D space and interpolated together obtaining the 3D shape of the finger (shape-from-silhouette). The 3D model is used to correct the perspective of the 5 original images, obtaining the corresponding ortho-images. Using the correlation between adjacent views, the five ortho-images are mosaicked, generating the first approximation of a rolled-equivalent fingerprint.

2.2.4 Single View Imaging by Using Time-of-Flight

3D Camera is also called Time-of-Flight (ToF) camera. Time-of-Flight technology, based on measuring the time that light emitted by an illumination unit requires to travel to an object and back to a detector, is used in Light Detection and Ranging scanners for high-precision distance measurements. Recently, this principle has been the basis for the development of new range-sensing devices, which are realized in standard CMOS or CCD technology.

The time-of-flight 3D laser scanner is an active scanner that uses laser light to probe the subject. At the heart of this type of scanner is a time-of-flight laser rangefinder. The laser rangefinder finds the distance of a surface by timing the round-trip time of a pulse of light. A laser is used to emit a pulse of light and the amount of time before the reflected light is seen by a detector is timed. Since the speed of light C is known, the round-trip time determines the travel distance of the light, which is twice the distance between the scanner and the surface. If t is the round-trip time, then distance is equal to $(C*t)/2$. The accuracy of a time-of-flight 3D laser scanner depends on how precisely we can measure the t time. Panasonic has developed a commercial ToF camera (D-IMager by Panasonic 2010).

There are two main approaches currently employed in ToF technology. The first one utilizes modulated, incoherent light, and is based on a phase measurement. The second one is based on an optical shutter technology. The main details about the technology will be showed in the chapters that introduce the 3D face verification system.

2.3   3D Biometrics Technologies

After the imaging technologies discussed, the feature extraction technologies will be presented in this chapter. Same to the imaging technologies, different systems use different feature extraction technologies. 3D feature extractions are different with the 2D's. But each has their own advantages and disadvantages, and they can be used both in the same systems. In the following sections, the feature extraction technologies will be expounded concisely and separately corresponding to the systems.

2.3.1   3D Ear Recognition

Iannarelli's research is widely known as the earliest work on using ear for person identification. By means of manually analyzing similarity among 10,000 ears from different persons, his experiment suggested that the uniqueness of human ear can be ensured based on a limited number of features or characteristics (Iannarelli 1989). Burge and Burger proposed an adjacency graph, which is built from the Voronoi diagram of the ear's edge segments, to describe the ear (Burge and Burger 2000). Chang et al. used principal component analysis to ear images (Chang et al. 2003). However 2D schema has its intrinsic limitation to grasp rounded surface details, especially spatial ridge and valley distribution.

In comparison, 3D ear holds rounded ear surface shape information. 3D ear model can be sampled in a relatively flexible way by range scanner device, e.g. Minolta Vivid 910 range scanner. It is more acceptable for relaxing people in sampling procedure. Chen and Bhanu presented a 3D ear recognition method using a local surface shape descriptor (Chen and Bhanu 2003). They also described a two-step ICP (Iterative Closest Point) procedure for matching 3D ears (Chen and Bhanu 2005), then they proposed two different representations for surface matching: The first representation is the ear helix/antihelix, whose 3D coordinates are obtained from the detection algorithm; the second representation is the local surface patch (LSP), which is invariant to rotation and translation (Chen and Bhanu 2007). Yan et al. has presented an ear biometrics system using 2D and 3D information. The automatic ear extraction algorithm can crop the ear region from the profile image, separating the ear from hair and earring. The recognition subsystem uses an ICP-based approach for 3D shape matching (Yan and Bowyer 2007).

Chen presented a general framework for efficient recognition of highly similar 3D objects which combines the feature embedding and SVM rank learning techniques. They use the UCR data set for testing, and get good performance (Chen and Bhanu 2009).

2.3.2 3D Palmprint Recognition

Compared with other 3D biometrics characteristics, 3D palmprint has some desirable properties. For instance, it is more convenient to collect and more user friendly; projecting stripes on palm has much higher acceptability than on face. One disadvantage of 3D palmprint may be that the palm surface is relatively plane so that the depth information of palm is more difficult to capture than that of face or ear. However, with the feature extraction and matching procedures in this book, the 3D palmprint recognition can reach very high accuracy.

Zhang et al. present a new personal authentication system that simultaneously exploits 2D and 3D palmprint features. The developed system uses an active stereo technique, structured light, to simultaneously capture 3D image or range data and a registered intensity image of the palm. The surface curvature feature based method is investigated for 3D palmprint feature extraction while Gabor feature based competitive coding scheme is used for 2D representation. The experiments on a database of 108 subjects achieved significant improvement in performance with the integration of 3D features as compared to the case when 2D palmprint features alone are employed. They also present experimental results to demonstrate that the biometrics system is extremely difficult to circumvent, as compared to the currently proposed palmprint authentication approaches (2010).

Li et al. proposed a scheme to jointly use palmprint 2D and 3D features for personal authentication in the feature level fusion. They extracted three levels of 2D and 3D palmprint features: shape features, principal line features and texture features. Then the ICP method is performed for alignment refinement to the texture feature map according to the principal line features and shape features when necessary. The alignment refinement reduces greatly the translation and rotation variations introduced in the palmprint data acquisition process. A couple of feature matching rules and an efficient joint 2D and 3D feature matching scheme were then proposed to fully use the 2D and 3D palmprint information. The experimental results on the 2D + 3D palmprint database, which contains 8,000 samples collected from 200 volunteers, demonstrated that the proposed method increases significantly the palmprint verification accuracy. Meanwhile, its average matching time is even less that by the classical weighting average fusing method (Li et al. 2010).

Figure 2.6 shows the 3-D palmprint acquisition system developed by the Biometrics Research Center, The Hong Kong Polytechnic University. There is a peg in the developed device serving as a control point to fix the hand. When the user puts his/her palm on the system, an LED projector will generate structured

Fig. 2.6 The 3D palmprint authentication system developed by the Biometrics Research Centre, The Hong Kong Polytechnic University

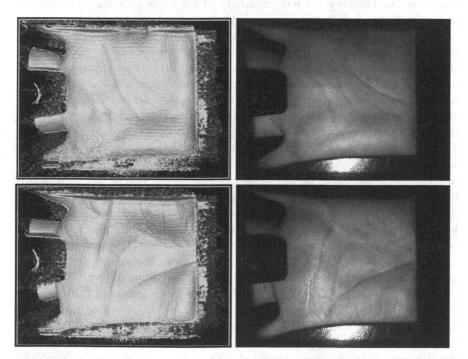

Fig. 2.7 3D palmprint samples and their corresponding 2D images

light stripes and project them to the palm. A series of gray-level images of the palm with the stripes on it is captured by the CCD camera, and then the depth information of the palm surface is reconstructed from the stripe images, some samples captured by this system are shown in Fig. 2.7 (Zhang 2009).

Cui and Xu extracted the 3D palmprint features using the appearance-based linear discriminant analysis (LDA) method. They also investigated the relationship between the recognition accuracy and the resolution of the 3D palmprint image. The experimental results show that the 3D palmprint images with resolution 16×16 and 32×32 are better for 3D palmprint recognition. The experiment results also show the feasibility of our method (Cui and Xu 2011).

2.3.3 3D Fingerprint Recognition

Compared with other biometrics features, fingerprint-based biometrics is the most proven technique and has the largest market shares. Traditional fingerprint system is based on the 2D fingerprint image. The system is not robust because the image captured is sensitive to many factors such as the change of the finger ridge structure due to injuries or heavy manual labors, data distortion under different illumination, uncontrollability and non-uniformity of the finger pressure on the device and so on (Wang et al. 2009).

In 3D fingerprint images, more features can be extracted. A coarser level of fingerprint features than level 1 features—Level Zero Features which refers to finger structural features are defined by Kumar and Zhou (2011). These features are then used for assisting fingerprint identification and gender classification. Experimental results show that an accuracy of 85.1 % can be achieved when using 3D curve-skeleton for recognition. The sectional curvatures can be used for human gender classification and an accuracy of 81 % is obtained in our database. An EER of 1.5 % is realized by including Level Zero Features into fingerprint recognition which demonstrates the effectiveness of 3D fingerprint recognition. Simple feature extraction and matching algorithm are used in the system. We believe that higher accuracy can be achieved if more advanced feature extraction and matching methods are proposed in the future. Discover the relationship between different levels of fingerprint features and propose more powerful fusion strategy will future improve 3D fingerprint recognition performance. Parziale et al. proposed a multi-camera touchless fingerprint scanner which acquires different finger views that are combined together to provide a 3-D representation of the fingerprint (Parziale et al. 2006; TBS 2007). Maev introduced a newer development of the ultrasonic fingerprint imaging using a scanning acoustic microscope to obtain images and acoustic data in the form of 3D data array. Important feature reveled on the acoustic images was the clear appearance of the sweat pores, which could provide additional means of identification (Maev et al. 2009). Yongchang Wang et al. employed a non-contact, 3-D scanning method of structured light illumination through phase measuring profilometry to make a 3-D scan of the human finger (Salvi et al. 2002; Yalla and Hassebrook 2005; Srinivasan et al. 1984; Wang et al. 2010).

Dr. Yongchang Wang from 3D Imaging Laboratory, University of Kentucky developed another touchless 3D fingerprint system, which can capture and process higher quality fingerprint data. The system relies on a real-time 3D sensor using structured light illumination (SLI), which generates both texture and detailed ridge depth information (Wang et al. 2009). TBS Inc. (TBS) developed the first 3D touchless fingerprint system in 2005 (TBS 2005).

2.3.4 3D Face Recognition

The traditional face recognition technology is usually based on two dimensional data. Though it was partially successful, often proves not to be robust under adverse recognition conditions. It is susceptible to errors caused by changes in lighting condition, expression, head pose and image capture quality. The 3D information of the face can eliminate some of the intrinsic problems associated with 2D recognition system. For instance, the 3D surface of a face is invariant to changes in lighting conditions (Bardsley 2005).

As the 3D face recognition developed rapidly, the 3D face feature extraction has attracted many interests (Gong 2006). To make the face recognition system fully automatic, facial feature extraction is one of the crucial steps (Lu and Jain 2005). Beumier and Acheroy designed a prototype for 3Dface recognition by using structured light. The 3D comparison is carried out by profile matching, either globally or more specifically for central and lateral profiles. A couple of seconds suffice to get a 3D representation and compare it to the claimed reference (Beumier and Acheroy 2000). Wang et al. focus on both 3D range data and 2D gray-level facial images. They extracted shape features from 3D feature points which are described by Point Signature in the 3D domain and texture features from 2D feature points which are described by Gabor filter responses in the 2D domain (Wang et al. 2002). Lu et al. focus on the key point feature extraction and select a subset of the facial landmarks as the feature points which include nose tip, two inner eye corners, two outside eye corners, and two mouth corners (Lu and Jain 2005). Gordon (Gordon 1992) extracted features which are classified into high level features and low level scalar features, from range and curvature data. Queirolo et al. proposed a method that uses a simulated annealing-based approach (SA) for range image registration with the surface interpenetration measure (SIM) as similarity measure. Comprehensive experiments were performed on the FRGC v2 database, a verification rate of 96.5 % was achieved at a FAR of 0.1 %. In the identification scenario, a rank-one accuracy of 98.4 % was achieved (Queirolo et al. 2010). Berretti et al. present an approach which takes into account geometrical information of the 3D face and encodes the relevant information into a compact representation in the form of a graph. The graph-based representation permits very efficient matching for face recognition. The method obtained the best ranking at the SHREC 2008 contest for 3D face recognition (Berretti et al. 2010). Kemelmacher-Shlizerman proposes a novel method for 3D shape recovery of faces, which obtains as input a single image and uses a mere single 3D reference model of a different person's face (Kemelmacher-Shlizerman and Basri 2011).

Since the 3D face recognition technology has been researched, there have been many systems about 3D face recognition and some systems have been implemented in the daily life. Such as the Hanvon, terminal manufacturer of 3D facial biometrics FaceID F710, distributed in Spain by Kimaldi, has carried out the installation of these terminals in China Citic Bank Asian country (Hanvon 3D facial recognition system 2012).

The purpose is to replace the traditional recognition system, identification of people and access control, by a 3D face recognition system which can ensure a high, rapid and non-contact recognition process.

2.4 Security Applications

Biometrics applications span a wide range of vertical markets, including security, financial/banking, access control, healthcare and government applications. Biometrics can be used in both customer and employee oriented applications such as ATMs, airports, and time attendance management with the goals of improving the workflows and eliminating fraud.

It is expected that the use of 3D identification systems to supplement or even replace existing services and methods in some applications with high security requirement, such as border control, Citizen ID Program, Banking, military etc.

2.4.1 Border Control

Passengers going aboard or entering a country must present passports and other border-crossing documents to the Border Guard. It will spend time for the border guard to verify these documents. In order to let border control becoming faster, convenient and safer, now there are more and more countries start using biometrics passport, such as United States, Canada, Australia, Japan, Hong Kong and so on. With the development of 3D biometrics technologies and system, we believe they will play an important role in border control for their high accuracy.

2.4.2 Citizen ID Program

It is a trend for governments to use biometrics technology on the issuance of citizen identity cards. In Hong Kong, a project called Smart Identity Card System (SMARTICS) uses the fingerprint as the authentication identifier. Efficient government services using SMARTICS will provide increased security and faster processing times on different operations such as driver license or border crossing. We think that 3D biometrics technologies are effective to be used on similar applications.

2.4.3 Banking

The internal operation of banking such as daily authentication process can be replaced by using biometrics technology. Some banks have implemented an authorization mechanism for different hierarchy of staff by swiping their badge for audit trail purpose. But a supervisor's badge may be stolen, loaned to other members of staff or even lost. Biometrics system eliminates these kinds of problems by placing an identification device on each supervisor's desk. When a junior member of staff has a request, it is transmitted to the supervisors' computer for biometrics approval and automatically recording.

2.4.4 Military

Department of Defense of USA distributed more than 11 million Common Access Cards (CAC) as its primary form of identification and enhanced protection to the military network. Although the CAC has proved to be a valuable tool, there are still security gap concerns if cards are lost or stolen and corresponding Personal Identification Numbers are cracked. To fill that void, the Air Force is using biometrics as a way to provide positive identification and authentication (Biometric Technology Working for Military Network 2008). Biometrics is also being used in support of the war on terrorism. Combined with other security measures, biometrics has fast become the preferred solution to military controlled access, can keep track of who has entered to particular areas because biometrics cannot be shared or borrowed.

2.5   Summary

In this chapter, the imaging technologies have been discussed separately. Each imaging technology has its advantages and disadvantages and fit different systems. Same to the imaging technology, the 3D feature extraction technologies and systems are also different. Thus, we have some preliminary understanding of the 3D biometrics recognition technologies. In the following chapters, the 3D ear recognition system, 3D palmprint recognition system, 3D face recognition system and 3D fingerprint recognition system will be presented separately.

References

Bardsley D, Li B (2005) Year 1 annual review stereo vision for 3D face recognition. Dissertation, University of Nottingham. doi:10.1.1.111.5769
Berretti S, Del Bimbo A, Pala P (2010) 3D face recognition using isogeodesic stripes. IEEE Trans Pattern Anal Mach Intell 32(12):2162–2177. doi:10.1109/TPAMI.20-10.43
Beumier C, Acheroy M (2000) Automatic 3D face authentication. Image Vis Comput 18(4):315–321. doi:10.1016/S0262-8856(99)00052-9

Biometric Technology Working for Military Network (2008). http://americancityandcou-nty.com/security/military-using-biometrics-0221

Burge M, Burger W (2000) Ear biometrics in computer vision. In: Proceedings of the international conference on pattern recognition, vol 2, pp 822–826. doi:10.1109/ICPR.2000.906202

Chang K, Bowyer KW, Sarkar S, Victor B (2003) Comparison and combination of ear and face images in appearance-based biometrics. IEEE Trans Pattern Anal Mach Intell 25(9):1160–1165. doi:10.1109/TPAMI.2003.1227990

Chen H, Bhanu B (2003) Human ear recognition in 3D. In: Proceedings of the workshop multimodal user authentication, pp 91–98

Chen H, Bhanu B (2005) Contour matching for 3D ear recognition. In: Proceedings of the seventh IEEE workshop on applications of computer vision. doi:10.1109/ACVM-OT.2005.38

Chen H, Bhanu B (2007) Human ear recognition in 3D. IEEE Trans Pattern Anal Mach Intell 29(4):718–737. doi:10.1109/TPAMI.2007.1005

Chen H, Bhanu B (2009) Efficient recognition of highly similar 3D objects in range images. IEEE Trans Pattern Anal Mach Intell 31(1):172–179. doi:10.1109/TPAMI.2008.176

Chen Y, Parziale G, Diaz-Santana E, Jain AK (2006) 3D touchless fingerprints: compatibility with legacy rolled images. Biometric consortium conference. biometrics symposium: special session on research. pp 1–6. doi:10.1109/BCC.2006.4341-621

Cui J, Xu Y (2011) Three dimensional palmprint recognition using linear discriminant analysis method. 2011 second international conference on innovations in bio-inspired computing and applications, pp 107–111. doi:10.1109/IBICA.2011.31

D-IMager by panasonic (2010). http://www2.panasonic.biz/es/densetsu/device/3DImageSenso-r/en/index.html

Gong X (2006) Automatic 3D face segmentation based on facial. IEEE international conference on industrial technology, pp 1154–1159. doi:10.1109/ICIT.2006.372409

Gordon GG (1992) Face recognition based on depth and curvature features. IEEE computer society conference on computer vision and pattern recognition (CVPR'92), pp 808–810. doi:10.1109/CVPR.1992.223253

Hung YY, Lin L, Shang HM, Park BG (2000) Practical threedimensional computer vision techniques for full-field surface measurement. Opt Eng 39(1):143–149. doi:10.1117/1.602345

Iannarelli A (1989) Ear identification, USA, ISBN: 0962317802

Judge TR, Bryanston-Cross PJ (1994) A review of phase unwrapping techniques in fringe analysis. Opt Lasers Eng 21(4):199–239. doi:10.1016/0143-8166(94)90073-6

Kanade T, Saito H, Vedula S (1998) The 3D room: digitizing time varying 3D events by synchronized multiple video streams. Technical report, CMU-RI-TR-98-34

Kemelmacher-Shlizerman I, Basri R (2011) 3D face reconstruction from a single image using a single reference face shape. IEEE Trans Pattern Anal Mach Intell 33(2):394–405. doi:10.1109/TPAMI.2010.63

Kumar A, Zhou Y (2011) Contactless fingerprint identification using level zero features. IEEE computer society conference on computer vision and pattern recognition workshops, pp 114–119. doi:10.1109/CVPRW.2011.5981823

Schirmacher HM Li, Seidel HP (2001) On-the-fly processing of generalized lumigraphs. In: Eurographics. doi:10.1111/1467-8659.00509

Li W, Zhang L, Zhang D, Lu G, Yan J (2010) Efficient joint 2D and 3D palmprint matching with alignment refinement. Comput Vis Pattern Recogn 795–801:0. doi:10.1109/CVPR.2010.5540134

Lu X, Jain A K (2005) Multimodal facial feature extraction for automatic 3D face recognition. Technical report MSU-CSE-05-22, Michigan State University. doi:10.1.1.92.8839

Maev RG, Bakulin EY, Maeva EY, Severin FM (2009) High resolution ultrasonic method for 3D fingerprint representation in biometrics. Acoust Imaging 29:279–285. doi:10.1007/978-1-4020-8823-0_39

McMillan L, Gortler S, Buehler C, Matusik W, Raskar R (2000) Image-based visual hulls. In: Proceedings of the 27th annual conference on computer graphics and interactive techniques, pp 369–374. doi:10.1145/344779.344951

Naemura T, Tago J, Harashima H (2002) Real-time video-based modeling and rendering of 3D scenes. IEEE Comput Graph Appl 22:66–73. doi:10.1109/38.9-88748

Parziale G, Diaz-Santana E, Hauke R (2006) The surround imager: a multi-camera touchless device to acquire 3D rolled-equivalent fingerprints. In: Proceedingsof IAPR international conference on biometrics, vol 3832, pp 244–250. doi:10.1007/11608288_33

Photon-X (2010). http://www.photon-x.com/3D_Biometrics.html

Queirolo CC, Silva L, Bellon ORP, Segundo MP (2010) 3D face recognition using simulated annealing and the surface interpenetration measure. IEEE Trans Pattern Anal Mach Intell 32(2):206–219. doi:10.1109/TPAMI.2-009.14

Salvi J, Armangue X, Batlle J (2002) A comparative review of camera calibrating methods with accuracy evaluation. Pattern Recogn 35(1):1617–1635. doi:10.1016/S0031-3203(01)00126-1

Srinivasan V, Liu H, Halioua M (1984) Automated phase-measuring profilometry of 3D diffuse objects. Appl Opt 23(18):3105–3108. doi:10.1364/AO.23.003105

Hanvon 3D facial recognition system (2012). http://www.kimaldi.com/kimaldi_eng/news/the_manufacturer_of_3D_hanvon_facial_recognition_biometrics_reaches_the_100_000_terminals_faceid_installed

TBS (2005). http://www.send2press.com/newswire/2005-04-0405-008.shtml

TBS (2007). http://www.tbs-biometrics.com/

Wang Y, Chua C-S, Ho YK (2002) Facial feature detection and face recognition from 2D and 3D images. Pattern Recogn Lett 23(10):1191–1202(12). doi:10.1016/S0167-8655(02)00066-1

Wang Y, Hao Q, Fatehpuria A, Hassebrook LG, Lau DL (2009) Data acquisition and quality analysis of 3-dimensional fingerprints. IEEE conference on biometrics, identity and security. doi:10.1109/BIDS.2009.5507527

Wang Y, Hassebrook LG, Lau DL (2010) Data acquisition and processing of 3D fingerprints. IEEE Trans Inf Forensics Secur 5(4):0. doi:10.1109/TIFS.2010.2062177

Wilburn BS, Smulski M, Lee K, Horowitz MA (2002) The light field video camera. In: Proceedings of media processors 2002, SPIE electronic imaging. doi:10.1.1.115.7800

Yalla V, Hassebrook LG (2005) Very-high resolution 3d surface scanning using multi-frequency phase measuring profilometry. Spaceborne sensors II, SPIE's defense and security symposium, vol 5798–09, pp 1234–1240. doi:10.1117/12.603832

Yan P, Bowyer KW (2007) Biometric recognition using 3D ear shape. IEEE Trans Pattern Anal Mach Intell 29(8):1297–1308. doi:10.1109/TP-AMI.2007.1067

Yang JC, Everett M, Buehler C, McMillan L (2002a) A real-time distributed light field camera. In: Eurographics workshop on rendering, pp 1–10. doi:10.1.1.4.882

Yang R, Welch G, Bishop G (2002b) Real-time consensus-based scene reconstruction using commodity graphics hardware. In: Proceedings of pacific graphics. doi:10.1111/1467-8659.00661

Zhang D (2004) Palmprint authentication. Norwell. ISBN: 1-4020-8096-4

Zhang D, Lu G, Li W, Zhang L, Luo N (2009) Palmprint recognition using 3-D information. IEEE Trans Syst Man Cybern 39(5):505–519. doi:10.1109/TSMCC.2009.2020790

Zhang D, Kanhangad V, Luo N, Kumar A (2010) Robust palmprint verification using 2D and 3D features. Pattern Recogn 43:358–368. doi:10.1016/j.patcog.2009.04.026

Part II
3D Ear Recognition Based on Line Structured Light

Chapter 3
3D Ear Acquisition System

Abstract The human ear is a new feature in biometrics that has several merits over the more common face, fingerprint and iris biometrics. Unlike the fingerprint and iris, it can be easily captured from a distance without a fully cooperative subject. Also, unlike a face, the ear has a relatively stable structure that does not change much with the age and facial expressions. Based on a special solid ear structure, in PART I, we explore all aspects of 3D ear recognition, such as representation, detection, recognition, indexing and performance evaluation. This Chapter presents a novel method of 3D ear acquisition system and 3D ear coordinate direction normalization based on projection density. It has been written for a professional audience of both researchers and practitioners within industry, and is also ideal as an informative text for graduate students in computer science and engineering.

Keywords 3D ear • Biometrics • Line structured-light imaging • 3D ear acquisation

3.1 Introduction

In today's complex, geographically mobile, increasingly electronically wired information society, the problem of verifying an individual identity continues to pose a great challenge. Conventional technology using Personal Identification Numbers (PIN) or passwords, often in conjunction with plastic cards, is neither convenient nor particularly secure. In the quest for a superior solution, biometrics verification techniques are fast emerging as the most reliable and practical method of individual identity verification (Zhang 2000; Jain 1999).

The ear has proven to be a stable candidate for biometrics authentication due to its desirable properties such as universality, uniqueness and permanence (Burge and Burger 2000). In addition, the ear has rich features: its structure does not change with the age; and its shape is not affected with facial expressions.

D. Zhang and G. Lu, *3D Biometrics*, DOI: 10.1007/978-1-4614-7400-5_3,
© Springer Science+Business Media New York 2013

Researchers have developed several approaches for ear recognition from 2D images (Burge and Burger 2000; Hurley et al. 2005; Liu and Yan 2008). Burge and Burger proposed a method based on Voronoi diagrams (Burge and Burger 2000). They built adjacency graphs from Voronoi diagrams and used a graph matching based algorithm for authentication. Hurley, Nixon and Carter developed a system based on force field feature extraction (Hurley et al. 2005). They treated the ear image as an array of mutually attracting particles that act as the source of a Gaussian force field. Choras presents a geometrical method of feature extraction from human ear images (Liu and Yan 2008). Although these approaches show some good results, the performances of 2D ear authentication will always affected by the illuminations and pose variation. Also, the ear has more spatial geometrical information than texture information, but spatial information such as posture, depth, and angle are limited in 2D ear images.

In recent years, 3D techniques have been used in biometrics authentication, such as 3D face (Kakadiaris et al. 2007; Samir et al. 2006), 3D palmprint (Zhang 2009) and 3D ear recognition (Chen and Bhanu 2007; Yan and Bowyer 2007; Chen and Bhanu 2009). A 3D ear image is robust to imaging conditions, and contains surface shape information which is related to the anatomical structure as well as being insensitive to environmental illuminations. Therefore, 3D ear recognition has drawn researchers' attention recently. Chen and Bhanu (2007) proposed a 3D ear recognition algorithm based on local surface patch and ICP method. They also presented a 3D ear indexing method (Chen and Bhanu 2009) which combined feature embedding and a support vector machine based learning technique for ranking the hypotheses. Yan and Bowyer (2007) presented an automated segmentation method by finding ear pit and using an active contour algorithm on both color and depth images, in addition to describing an improved ICP approach for 3D ear shape matching.

Most of the previous researchers use commercial laser scanners to acquire the 3D range image, for example, the widely used Minolta VIVID Series. However the scanners are always expensive and inconvenient to assemble complete systems for real applications. With this consideration, we developed a low-cost laser scanner specifically designed for online 3D ear acquisition using the laser-triangulation principle. This chapter proposes a framework to acquire high quality 3D ear images in real time. It uses a CCD camera and laser projector as the core components to reconstruct the 3D ear image. The system is designed to have the best performance at a reasonable price so that it is suitable for civilian personal identification applications.

3.2 Capturing Device Design

3.2.1 The Principle of Triangulation Imaging

Structured-light imaging method is a widely used 3D imaging method as it is highly accurate. A projector projects a certain pattern of structured light on the surface. The structured light is modulated by the surface shape, and the

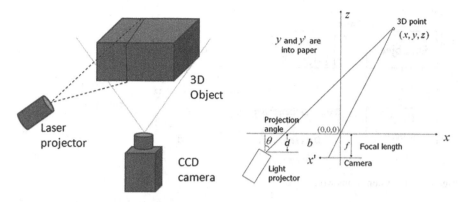

Fig. 3.1 Imaging principle of the laser-triangulation technique

modulated stripes are captured by a CCD camera which has a fixed distance from the projector. According to the modulated stripe images and the geometric correlation between the measured surface, projector and CCD camera, the distance from the measured surface to the reference plane can be calculated. Laser-triangulation imaging method is one of the line structured light imaging methods which is widely used due to its high accuracy and speed. We use the laser-triangulation principle to develop a low-cost laser scanner for ear acquisition. Figure 3.1 illustrates the imaging principle of the laser-triangulation technique.

In Fig. 3.1 the laser projector projects laser light on the object at angle θ. On the surface of the object, the laser light is shown as red lines modulated by the surface shape. The CCD camera captures the 2D object image with laser lines on it. Then the 3D coordinates of points on the laser lines can be calculated.

3.2.2 System Framework

Figure 3.2 illustrates the 3D ear data collection and processing steps. The computer controls a projector to project a series of laser-light stripes on the ear surface, and the CCD camera captures the ear images with projected stripes on it. At the same time the computer sends a command to the data collection board to store the images. The data collection takes about 2 s. From these ear images, the depth information of each point on the ear can be computed using laser-triangulation techniques. These steps which are marked using blue arrows in Fig. 3.2, will take around 1 s. Therefore, the total time for 3D ear generation is about 3 s.

There are mainly 4 components of the 3D laser scanner (refer to Fig. 3.3): Motor driver and controller Fig. 3.3a, Motor and laser projector Fig. 3.3b, Camera Fig. 3.3c, and Computer with image capturing card Fig. 3.3e.

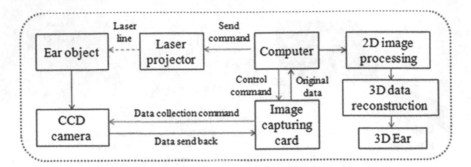

Fig. 3.2 The system framework

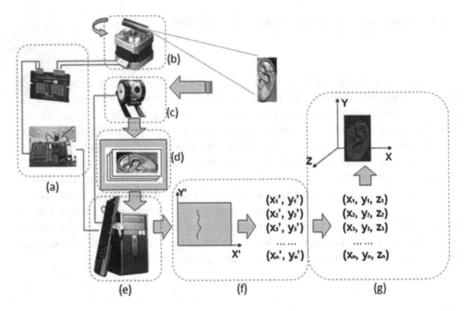

Fig. 3.3 Components of the system. **a** Motor driver and controller. **b** Motor and laser projector. **c** Camera. **d** A serial of 2D ear images with laser lines. **e** Computer. **f** Coordinates of the laser line points. **g** 3D ear

When the device scans an ear:

1. The computer Fig. 3.3e send a command to the control circuit Fig. 3.3a, and the motor driver Fig. 3.3a makes the step motor Fig. 3.3b rotate in a preset speed and direction. The motor puts the projector Fig. 3.3b in motion at the same speed and direction, and projects laser lines on the captured ear.
2. At the same time as step (1), the computer sends a command via the image capturing card Fig. 3.3e and enables the CCD camera to Fig. 3.3c capture the ear images with laser lines on them.
3. Then, a serial of 2D ear images with laser lines Fig. 3.3d are sent back to the computer Fig. 3.3e.

4. Afterwards, we extract the laser lines in each 2D image and obtain a series of 2D coordinates of the laser lines points Fig. 3.3f.
5. Using the laser-triangulation principle we translate the 2D coordinates to 3D, and obtain the 3D image of the ear Fig. 3.3g.

3.2.3 System Calibration

Calibration is an important step for 3D object measurement. Here, we use a comprehensive calibration method. There are three parameters b, d and f which need to be decided by calibration as shown in Fig. 3.4, where b and d are horizontal and vertical offset between the motor shaft and the camera's optical center, while f is the focal length of the camera. According to the following equations,

$$\vec{p} = \begin{pmatrix} x \\ y \\ z \end{pmatrix} = \begin{pmatrix} \frac{x'(b-d\tan\theta)}{x'+f\tan\theta} \\ \frac{y'(b-d\tan\theta)}{x'+f\tan\theta} \\ \frac{f'(b-d\tan\theta)}{x'+f\tan\theta} \end{pmatrix}. \tag{3.1}$$

We get:

$$z = \frac{f'(b-d\tan\theta)}{x'+f\tan\theta}. \tag{3.2}$$

We can see that if z, x' and θ are known, the unknown parameters are b, d and f. Actually we can calculate x' by the collected images, and obtain θ by the motor driver. Hence, given three different distances z_1, z_2, z_3, as shown in Fig. 3.4, we obtain three equations from Eq. (3.2), and the unknown parameters b, d and f can be resolved.

Fig. 3.4 Illustration of three distances for calibration

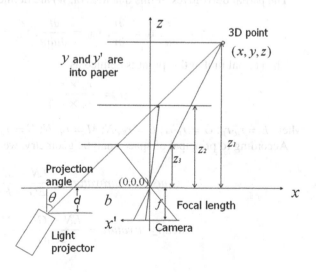

3.3 Ear Segmentation Based on Compound Curvature

Segmentation is an important step before coordinate direction normalization. Compound curvature efficiently helps to locate regions of interest of ear surface. Compound curvature is the combination of the Gaussian curvature and mean curvature, which represent intrinsic and extrinsic properties of surfaces, respectively.

For a given point on the surface, the principal curvatures k_1 and k_2 represent the degrees of maximum and minimum surface spread in the neighborhood of such a point. The Gaussian curvature, which actually is an intrinsic property of the surface, is equal to k_1k_2. Hence it does not depend on the particular embedding of the surface. However Gaussian curvature determines whether a surface is locally convex (when it is positive) or locally saddle (when it is negative). It can be used to detect ridges and valleys of a 3D model. The mean curvature is defined as $(k_1 + k_2)/2$. Unlike Gaussian curvature, it is extrinsic and depends on the embedding. Therefore, it can be used to measure the degree of embedding and to help us to segment the ear surface from background.

Scatter point cloud is the simplest data mode to construct a 3D model. It is easy to acquire using 3D imaging devices and is relatively lightweight for storing. The segmentation introduced below is based on 3D scattered points. Thus, this schema does not rely on meshing. We introduce a curvature evaluation method through local surface approximation. Then thresholding is applied for segmentation.

For a given point P, define a local neighborhood as the index set N_P. A parametric quadric is approximated by the index set N_P. Suppose this quadric is defined as

$$r = r(u, w). \tag{3.3}$$

According to surface quadric estimation method (Turk and Levoy 1994), r can be fitted. See Fig. 3.5: the Euclidean distance from k-neighbors to parametric quadric.

The partial derivatives of this quadric $r(u, w)$ are defined as follows

$$r_u = \frac{\partial r}{\partial u}, \ r_v = \frac{\partial r}{\partial v}, \ r_{uu} = \frac{\partial r}{\partial u \partial u}, \ r_{uv} = \frac{\partial r}{\partial u \partial v}. \tag{3.4}$$

The normal unit at this point is defined as

$$N = \frac{r_u \times r_v}{\|r_u \times r_v\|}, \tag{3.5}$$

where $E = r_u \cdot r_u; \ G = r_u \cdot r_v; \ L = r_{uu} \cdot N; \ M = r_{uv} \cdot N; \ N = r_{vv} \cdot N$.

According to principal of space analytic geometry, we have

$$Gaussian\_curavature = \frac{LN - M^2}{EG - F^2}, \tag{3.6}$$

$$Mean\_curvature = \frac{EN + GL - 2FM}{2(EG - F^2)}. \tag{3.7}$$

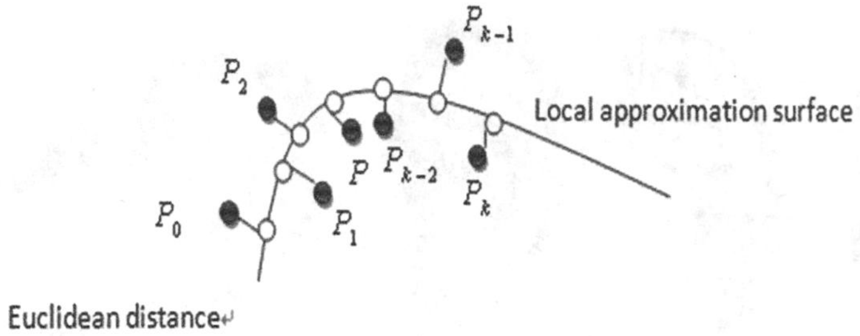

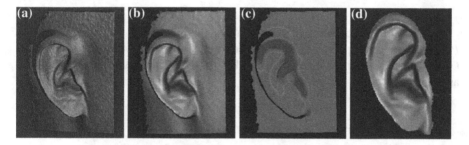

Euclidean distance↵

Fig. 3.5 Distances in the developed surface approximation

Fig. 3.6 Ear segmentation based on compound curvature. **a** Original 3D profile model.
b Gaussian smooth. **c** Curvature approximation. **d** Ear after segmentation

Curvature estimation is sensitive to noise. For stable curvature measurement,
we have to smooth the surface without losing the ear pit feature. Gaussian smoothing is applied on the data points. Then we apply thresholding to segment inner side
of ear from cheek, while close region determination helps to separates outer side
from skin of neck. Figure 3.6 gives an example of the smooth and segmentation
procedure.

3.4   Posture Normalization Method Using Projection Density

There are specific computable projective directions in 3D model, in which points
projected to the low dimensional spaces distributed most sparsely. Optimal projection pursuit is utilized to retrieve such directions. Project pursuit can be used to
project high dimensional data to low dimensional space and find the optimum projection vector of data in one dimension space by data character of research units.
The projective points on the XOY plane demonstrate special projection density, as
illustrated in Fig. 3.7.

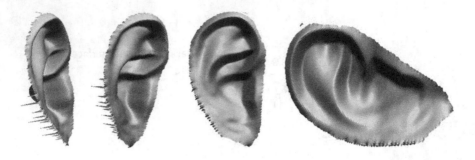

Fig. 3.7   3D ear model of different vertex projection direction

Suppose the projective points are P. After normalization, P is transformed to $P(x, y)$. For a given point $p_i(x, y)$ in the set, define its nearest k neighbors as NEI_i (x, y), which is a point cluster including k points. The neighboring point is noted as below

$$nei_j(x, y) \in NEI_i(x, y), j \in [1, k]. \tag{3.8}$$

Then define the projection density as

$$projection\_density = \frac{1}{N} \sum_{i=1}^{N} \sum_{j=1}^{k} dist(p_i(x, y), nei_j(x, y)), \tag{3.9}$$

where $dist(a, b)$ means the Euclidean distance between point a and b. Explained above, the projection density of certain projective point set actually implicitly relies on the angel of rotation on X and Y axes, as a result, projection density can be also expressed as $projection\_density = func(\alpha_x, \beta_y)$.

For certain projection direction, exhaustive calculation of projection density to given 3D ear model is implemented to analyze the data distribution characteristic of projection density. We calculate the projection density from -180 to $+180$ both along angle of rotation on X and Y axes. The result shows in Fig. 3.8.

In order to climb up to the peak value of the solution space, Newton descent method is applied:

$$\begin{aligned} \alpha_x(k+1) &= \alpha_x(k) - \eta \times \frac{\partial func(\alpha_x, \beta_y)}{\partial \alpha_x} \\ \beta_y(k+1) &= \beta_y(k) - \eta \times \frac{\partial func(\alpha_x, \beta_y)}{\partial \beta_y}, \end{aligned} \tag{3.10}$$

where $\alpha_x(k)$ is the angle of rotation on the X axes in kth iteration, and similarly for $\beta_y(k)$ on the Y axes. The learning rate η is assigned through experience. According to Eq. (3.10), $func(\alpha_x, \beta_y)$ is implicit continuously differentiable. Thus Newton descent method should be modified to satisfy discrete data. We introduce

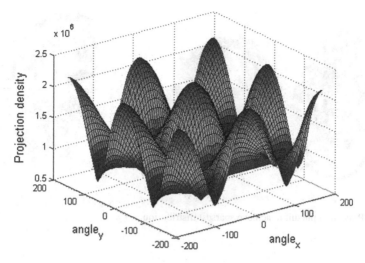

Fig. 3.8 Exhaustive calculation of projection density to certain 3D

a practicable method which does not rely on learning rate. The structure KD-tree (Bentley 1975) simplifies the procedure to search for K nearest neighbors, which is put into use for the algorithm. This method is described as below.

1. Construct a KD-tree for projective points in the *XYZ* plane. For each point, search for its *k* nearest neighbors. According to Eq. (3.10), calculate $func(\alpha_x, \beta_y)$. Similarly, step forward density along the X axes $func(\alpha_x + step_x, \beta_y)$ and step back density $func(\alpha_x - step_x, \beta_y)$ can also be worked out. $step_x$ is the value of the step length along the clockwise rotation in the X axes.
2. On the condition that $func(\alpha_x - step_x, \beta_y) < func(\alpha_x, \beta_y) < func(\alpha_x + step_x, \beta_y)$, the rotation angle α_x in X axes increases by step value $\alpha_x = \alpha_x + step_x$, and when $func(\alpha_x - step_x, \beta_y) > func(\alpha_x, \beta_y) > func(\alpha_x + step_x, \beta_y)$, reduce α_x by the step value: $\alpha_x = \alpha_x - step_x$. Except for two cases above, if $func(\alpha_x, \beta_y) > func(\alpha_x + step_x, \beta_y)$ and simultaneously $func(\alpha_x, \beta_y) > func(\alpha_x - step_x, \beta_y)$, α_x does not change its value while the step value $step_x$ is shortened twice.
3. The similar procedure for $func(\alpha_x, \beta_y)$, $func(\alpha_x, \beta_y - step_y)$, and $func(\alpha_x, \beta_y + step_y)$ is repeated. Update the value of β_y and $step_y$. Go back to the second step.
4. When $step_x$ and $step_y$ decrease below certain a threshold, the iteration terminates.

Based on this method, parameters α_x and β_y are fixed to ensure the sparsest status. Afterward γ_z: (angle of rotation on z axes) is the only parameter left to revise direction on the projection plane; in addition, some uncertainties need to be eliminated. Figure 3.9 shows the result of Newton descent method.

From Fig. 3.9, the maximum projection density is preserved in all three ear models; however, the obverse side and reverse side of ear have to be distinguished and we should rotate certain angle on the Z axes to ensure every ear model is laid

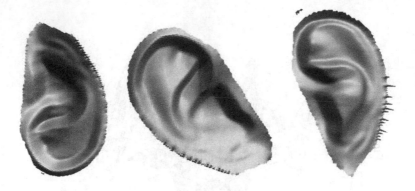

Fig. 3.9 Projection pursuit to achieve sparsest distribution

(a)   **(b)**   **(c)**

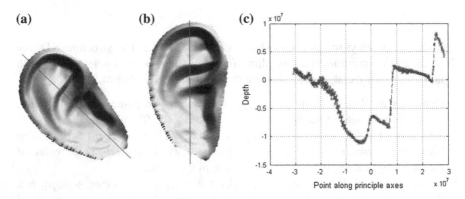

Fig. 3.10 Posture adjustment in 2D projection plane. **a** Sparest projection of ear. **b** Adjust result through PCA. **c** Points' depth along the principal axes

on uniform position in terms of specific rules. PCA's statistical property helps to do this work.

After the normalization of projective points on XOY plane, we calculate eigenvalues and eigenvectors of the covariance matrix of this projective point set. The eigenvector corresponding to the maximum eigenvalue determines the principal axis of the projective points, which is used to normalize the angle of rotation on Z axes. See Fig. 3.10. From (a) to (b), it shows clockwise rotation of the principal axes to vertical direction.

Next, distinguish the obverse side and reverse side by analyzing point depth distribution along the principal axes. Figure 3.10c shows this kind of distribution. Along the principal axes, divide points into k segments. For certain segments Seg_i, there are N_i points in this segment. Introduce a characteristic vector Dep $(d_1, d_2, d_3, \ldots, d_{k-2}, d_{k-1}, d_k)$ from which

$$d_i = \frac{1}{N_i} \sum_{p(x,y,z) \in Seg_i} p(z).$$

(3.11)

This characteristic vector can be used to efficiently detect surface visibility. Only define a group of standard vectors that corresponds to given postures according to prior knowledge, and then apply 1-NN. Compare the characteristic vector derived from testing model with these standard vectors, the reverse and obverse side can be discriminated. Figure 3.11 illustrates the final result. Ultimately, Ear of different postures are laid uniformly.

3.5 Experimental Results

There have been 250 individuals to collect their ear images using our 3D ear laser scanner, including 178 males and 72 females. We collected the 3D ears on two separate occasions, at an interval of around one month. On each occasion, the subject was asked to provide two left side face images and two right side face images. Therefore, each person provided 8 images, so that our database contained a total of 2,000 images from 500 different ears.

Figure 3.12 is a typical 3D ear sample captured by our device, where the top part is the 3D points cloud viewed at different viewpoints. We can see the 3D data cloud is smooth and able to display geometric shapes of the 3D ear. For testing the quality of the 3D ear sample, we computed the Gaussian curvature of the 3D points cloud. This result is shown in the low part in Fig. 3.12. It is clear that our 3D data could produce a good quality representation of the 3D ear.

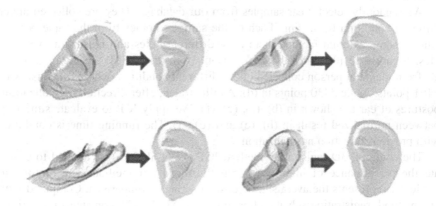

Fig. 3.11 Results after uncertainty elimination

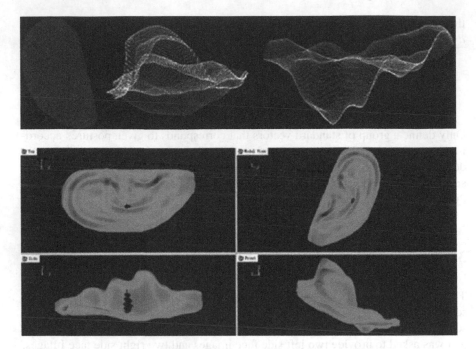

Fig. 3.12 Typical 3D ear sample captured by our device

Here we compare the proposed method with improved Mesh PCA. Mesh PCA is improved in (Besl and McKay 1992) to solve the problem of direction disorder in the adjustment. It detects the 3D surface visibility before coordinate transformation based on the sum of all normal vectors and introduces six different forms of rotation matrix to fix principal direction; however it is still a completely linear method based on statistic, which is sensitive to noise and relies on points' distribution on surface.

We randomly select 2 ear samples from our database. They are collected under different condition for testing. Each of the samples comes from the same person. Due to some external factors, there exist different holes and noises on two samples. Figure 3.13 illustrates the comparison. Both (a) and (d) are the input models from identical person collected under different condition, where (a) consists of 5461 points, while 5430 points in (d). As illustrated, after direction normalization, postures of ear are shown in (b), (c), (e), (f). We apply ICP to evaluate similarity between normalized result in (b), (e) and (c), (f). The running time is combined with previous direction normalization.

The common 3D registration method ICP (Li et al. 2010) is applied to evaluate the performance of improved Mesh PCA and our method. The runtime in Table 3.1 represents the average time cost in a single comparison. Compared with our method, registration schema after improved Mesh PCA consumes more time,

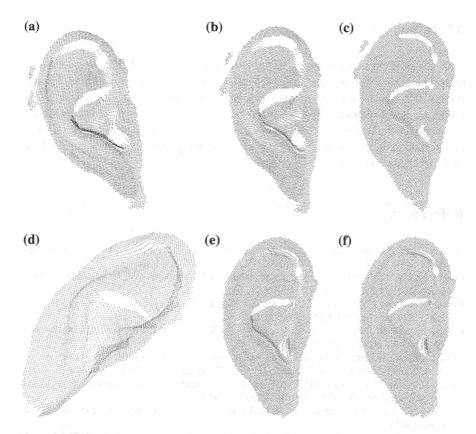

Fig. 3.13 Comparison between two methods **a** and **d** are input models from same person, **b** and **e** are corresponding results of improved Mesh PCA, and while **c** and **f** represent results of our method

Table 3.1 Comparison between our method and improved Mesh PCA

	Runtime (s)	Accuracy (%)
Our method	4	99.06
Improved Mesh PCA (Besl and McKay 1992)	15	98.03

which suggests distinct difference between two normalization results. Although ICP devotes to continuously decreasing model variance through rigid transformation, the accuracy of improved Mesh PCA is still lower than our method. In comparison, using method proposed in this chapter, posture is relatively unified after direction normalization. Thus, ICP quickly finds out corresponding point pairs and terminates iteration.

3.6 Summary

This Chapter presents a novel 3D ear imaging system and 3D ear coordinate direction normalization based on projection density. Compared with traditional methods, it has a lot of advantages that traditional methods do not possess. 3D modeling is not required to be organized in a mesh grid. Point cloud is applicable. It eliminates the uncertainty which exists in Mesh PCA and is robust to noises and holes. It generates relatively uniform posture, which provides convenience for feature extraction and speeds up registration.

References

Bentley JL (1975) Multidimensional binary search trees used for associative searching. Commun ACM 18(9). doi:10.1145/361002.361007

Besl P, McKay ND (1992) A method for registration of 3-D shapes. IEEE Trans Pattern Anal Mach Intell 14:239–256. doi:10.1109/34.121791

Burge M, Burger W (2000) Ear biometrics in computer vision. Proc Int Conf Pattern Recogn 2:822–826. doi:10.1109/ICPR.2000.906202

Chen H, Bhanu B (2007) Human ear recognition in 3D. IEEE Trans Pattern Anal Mach Intell 29(4):718–737. doi:10.1109/TPAMI.2007.1005

Chen H, Bhanu B (2009) Efficient recognition of highly similar 3D objects in range images. IEEE Trans Pattern Anal Mach Intell 31(1):172–179. doi:10.1109/TPAMI.2008.176

Hurley D, Nixon M, Carter J (2005) Force field feature extraction for ear biometrics. Comput Vis Image Underst 98:491–512. doi:10.1016/j.cviu.2004.11.001

Jain A (1999) Biometrics: personal identification in network society. Kluwer Academic, ISBN: 0792383451

Kakadiarisv IA, Passalis G, Toderici G, Murtuza MN, Lu YL, Karampatziakis N, Theoharis T (2007) Three-dimensional face recognition in the presence of facial expressions: an annotated deformable model approach. IEEE Trans Pattern Anal Mach Intell 29(4):640–649. doi:10.1109/TPAMI.2007.1017

Liu H, Yan J (2008) Multi-view ear shape feature extraction and reconstruction. 3rd international IEEE conference on signal-image technologies and internet-based system, pp 652–658. doi:10.1109/SITIS.2007.42

Li W, Zhang L, Zhang D, Lu G (2010) Efficient joint 2D and 3D palmprint matching with alignment refinement. In: Proceedings of IEEE international conference on computer vision and pattern recognition, pp 795–801. doi:10.1109/CVPR.2010.5540134

Samir C, Srivastava A, Daoudi M (2006) Three-dimensional face recognition using shapes of facial curves. IEEE Trans Pattern Anal Mach Intell 28(11):1858–1863. doi:10.1109/TPAMI.2006.235

Turk G, Levoy M (1994) Zippered polygon meshes from range images. In: Proceeding of conference on computer graphics and interactive techniques, pp 311–318. doi:10.1145/192161.192241

Yan P, Bowyer KW (2007) Biometric recognition using 3D ear shape. IEEE Trans Pattern Anal Mach Intell 29(8):1297–1308. doi:10.1109/TPAMI.2007.1067

Zhang D, Lu GM, Li W, Zhang L, Luo N (2009) Palmprint recognition using 3-D information. IEEE Trans Syst Man Cybern C Appl Rev 39(5):505–519. doi:10.1109/TSMCC.2009.2020790

Zhang D (2000) Automated biometrics: technologies and systems. Kluwer Academic Publishers, ISBN: 0-7923-7856-3

Chapter 4
Two Significant Characteristics in 3D Ear

Abstract This chapter introduces the methods to extract two significant character-istics in 3D ear images: ear-cheek angle and the dissimilarity of the left and right ears of the same person. The ear-cheek angle is defined as the angle between the normal vector of the cheek-plane and the normal vector of the ear-plane. To meas-ure the difference of the left and right ears a modified ICP (Iterative Closest Point) algorithm is applied. These characteristics cannot be extracted from 2D ear images. The experimental results show that the ear-cheek angle is unique and stable for each person and distinguishable for different people, which is a useful feature in 3D ear classification and indexing. In addition, the statistical analysis shows that there is an important difference between the left and right ears of the same person; hence both ears can be treated as a respective class in 3D ear recognition.

Keywords 3D ear recognition • Biometrics • Ear-cheek angle

4.1 Introduction

Biometrics authentication is playing important roles in applications of public secu-rity like access control (Pang et al. 2011), forensics, and e-banking (Zhang 2000; Jain 1999). In order to meet the needs of different security requirements, new biometrics including palmprint (Zhang et al. 2003), vein (Zhang et al. 2007) and so on, have been developed. The ear has proven to be a stable candidate for biom-etrics authentication due to its desirable properties such as universality, uniqueness and permanence (Burge and Burger 2000; Purkait and Singh 2008). In addition, an ear possesses several advantages: its structure does not change with age and its shape is not affected by facial expressions.

Researchers have developed several approaches for ear recognition from 2D images (Burge and Burger 2000; Hurley et al. 2005; Choras 2005). Burge and Burger proposed a method based on Voronoi diagrams (Burge and Burger 2000).

D. Zhang and G. Lu, *3D Biometrics*, DOI: 10.1007/978-1-4614-7400-5_4,
© Springer Science+Business Media New York 2013

They built an adjacency graph from Voronoi diagrams and used a graph matching based algorithm for authentication. Hurley, Nixon and Carter proposed a system based on force field feature: extraction (Hurley et al. 2005). They treated the ear image as an array of mutually attracting particles that act as the source of Gaussian force field. Choras presented a geometrical method of feature: extraction from human ear images (Choras 2005). Although these approaches show some good results, the performance of 2D ear authentication will always be marred by illuminations and pose variation. Also, the ear has more spatial geometrical information than texture information, but spatial information such as posture, depth, and angle are limited in 2D ear images.

In recent years, 3D techniques have been used in biometrics authentication, such as 3D face (Kakadiaris et al. 2007; Samir et al. 2006), 3D palmprint (Zhang et al. 2009) and 3D ear recognition (Bhanu and Chen 2007; Yan and Bowyer 2007; Chen and Bhanu 2009). A 3D ear image is robust to imaging conditions and contains surface shape: information which is related to the anatomical structure and insensitive to environmental illuminations. Therefore, 3D ear recognition has drawn the attention of more researchers recently. Chen and Bhanu proposed a 3D ear recognition method based on local: surface patch (Bhanu and Chen 2007), as well as a 3D ear indexing method which combined feature embedding and a support vector machine based learning technique for ranking the hypotheses (Chen and Bhanu 2009). Yan and Bowyer (2007) designed an automated segmentation method by finding the ear pit and using an active contour algorithm on both color and depth: images. They also described an improved ICP (Iterative Closest Point) approach for 3D ear shape matching. To date however, there has been no work with 3D ears that has extracted angle features between ear and cheek. The angle feature is stable for each person and is unique in 3D images. This characteristic gives 3D ear more special features than 2D ear, which is a helpful candidate for ear classification and indexing. As for the use of 3D techniques in ear biometrics, more features can be extracted from a 3D ear image. Hence, the difference between the left and right ears can be analyzed more specifically. We can then find out whether the left/right ears should be treated as the same ear or two different ears in biometrics authentication.

In this chapter, we propose a practical 3D ear recognition system using a laser 3D scanner developed by Hong Kong Polytechnic University (PolyU). Most of the previous work use commercial laser scanners to acquire the 3D range image, for example, the widely used Minolta VIVID series. However, these scanners are always expensive and not convenient to be assembled as part of a complete system for real applications. With this consideration, we developed a low-cost laser scanner for 3D ear acquisition using the laser-triangulation principle (as shown in Fig. 4.1a). The main components of our 3D ear scanner are CCD camera, laser projector, step-motor, and motion control circuit. 3D ear images are obtained as shown in Fig. 4.1b–c. For ear segmentation we used the relative position of endpoints in the laser lines to trace the ear boundary. After ear segmentation some ear edge points are used as references to locate a circular area on the cheek. Using these cheek and ear edge points, the hypothetic cheek-plane and ear-plane can

(a)   **(b)**   **(c)**

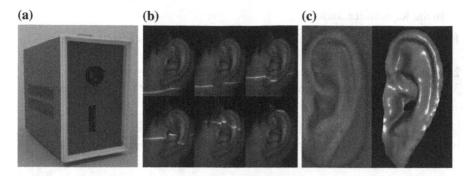

Fig. 4.1 A 3D ear recognition system developed in the biometrics research centre in Hong Kong
a 3D laser scanner, **b** and **c** are examples of ear acquisition

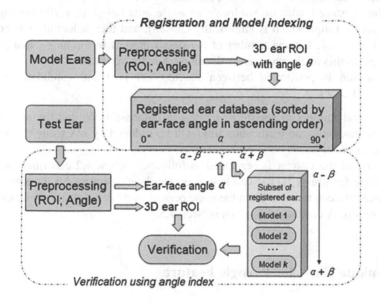

Fig. 4.2 The registration and model indexing diagram

be constructed, and subsequently the ear-cheek angle is quantitatively calculated.
Since this angle is stable for every person and distinguishable for different people,
it is a significant characteristic in 3D ear classification. Additionally, we compared
the differences between the left and right ears of the same person, and gave a statistical analysis to prove the distinguishable characteristics of an individual's two ears.

The primary function of the angle feature is to reduce the time in recognition. We
can use this feature to sort the samples in a registered database. When recognition is
performed we only need to compare the test ear with candidate ears, and determine
which has a closer angle value to the test ear. By using this method we do not need
to compare all the registered ears with the test ear, and thus a lot of time can be
saved. The registration and model indexing diagram is illustrated in Fig. 4.2.

In the Registration stage:

1. Given a model ear, we first locate the ROI which contains both the 3D ear and a square patch of the face.
2. Compute the angle θ between face plane and ear boundary plane.
3. According to the angle θ insert the ROI of the model ear into the registered ear database which is sorted by ear-cheek angle in ascending order.

Thus, the registered ears in database can be indexed by its angle, and the database stores both the 3D ears and their corresponding angle values.

In the Verification stage:

1. Given a test ear the location of the ROI which contains both the 3D ear (C_t) and a square patch of face is found.
2. Compute the angle α between face plane and ear boundary plane.
3. In the registered database we locate an angle area $[\alpha-\beta, \alpha + \beta]$ (the angle β is a search range which is empirically chosen), and then select all ear models (C_i, $i = 1,...,k$, k is the number of registered models in the angle area $[\alpha-\beta, \alpha + \beta]$) in this angle range as candidate models for verification.
4. Verification is performed between the test ear (C_t) and candidate models (C_i, $i = 1,...,k$).

If the total number of registered models in database is N, and the number of candidate models k, the verification time will be reduced by k/N using this angle-indexing method.

The rest of the chapter is organized as follows. Section 4.2 describes the ear-cheek angle feature in 3D ear and how it is calculated. Section 4.3 introduces the difference between two ears from the same person, and Sect. 4.4 shows our experimental results. A conclusion is given in Sect. 4.5.

4.2 Unique Ear-Cheek Angle Feature

This section describes a unique ear-cheek angle feature (shown in Fig. 4.3). In Sect. 4.2.1 we explain how it is calculated using two planes from both ear and cheek. These two planes are formed by first locating a Region of Interest (ROI) illustrated in Sect. 4.2.2, where both ear and cheek points are extracted and Principal Component Analysis (PCA) (Pang et al. 2010a, b) is applied to locate their normal vectors.

4.2.1 Definition

From Fig. 4.3a we can observe that there is an angle (θ) between the ear and cheek of a person. We assume there is a plane, $A_f x + B_f y + C_f z + D_f = 0$, which represents the 3D points on the cheek (yellow circle shown in Fig. 4.3a

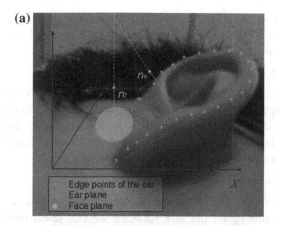

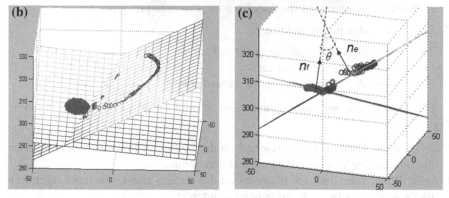

Fig. 4.3 Ear-cheek angle illustration. **a** Spatial relationship between ear and cheek, **b** ear/cheek plane, and **c** angle between ear and cheek plane

as well as the green circle shown in Fig. 4.3b–c). We also use another plane, $A_e x + B_e y + C_e z + D_e = 0$, to represent the 3D points on the ear edge (i.e., light blue square shown in Fig. 4.3a and dark blue dots shown in Fig. 4.3b–c). Thus, the normal vector of the cheek plane can be obtained as $n_f = (A_f, B_f, C_f)$, and the normal vector of the ear plane is $n_e = (A_e, B_e, C_e)$. The angle between cheek and ear can be defined as follows:

Let $\theta_l = \cos^{-1} \frac{n_f \cdot n_e}{|n_f| \times |n_e|}$, then

$$\theta = \begin{cases} \theta_l & \text{if } \theta_l < 90° \text{ as the ear - cheek angle} \\ 180° - \theta_l & \text{else.} \end{cases} \tag{4.1}$$

4.2.2 Angle Feature: Extraction

In order to calculate the ear-cheek angle, the ear is first located using a mask. Next, the laser lines on the ear are extracted and traced to represent the ear edge-points. Afterwards, ROI consisting of the ear and cheek can be defined. Finally, the normal vectors n_f and n_e are obtained from the located ear and cheek. Below we explain each step in detail.

1. Generating a mask:

   First, we select the last frame from the original images to acquire a binary image. Morphological dilation, opening, and closing options are then applied to fill holes and remove noisy pixels in this binary image. Lastly, the connected component labeling algorithm is used, and the largest connected component is retained as the final mask (as shown in Fig. 4.4a).

2. Extracting laser lines on the mask region:

   Since pixels on the laser lines are much brighter than others in the image (due to the laser shining on the skin surface), we can select the brightest pixel in each column to extract as the laser lines (as shown in Fig. 4.4b).

3. Tracing the edge points on the ear:

   This processing consists of two major tasks:
   (a) Tracing edge points on the ear.
   (b) Tracing edge points between the ear and cheek.

   For the first task, we can see three cases of points on the ear (marked in red in Fig. 4.5):

Case 1: When laser lines are projected on the bottom of the ear (earlobe), select the furthest line's right endpoint as an edge-point.
Case 2: When laser lines are projected on the ear and cheek select the rightmost line's right endpoint as the edge-point of interest.
Case 3: When there is only one line on the ear the rightmost endpoint is the edge-point we are interested in.

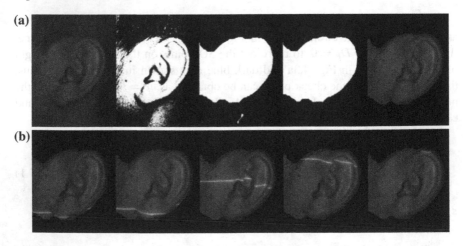

Fig. 4.4 Ear image processing **a** Ear mask generation, and **b** extraction of *laser lines* on the mask

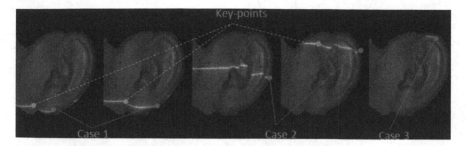

Fig. 4.5   Three cases of tracing the edge points on the ear with key-points

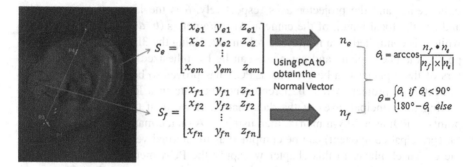

Fig. 4.6   Process of using ROI edge points (both ear and cheek) to calculate the ear-cheek angle

For every 2D image each edge-point found is set to its corresponding 3D coordinate and placed into Se (set of edge-points of the ear).

In the second task, key-points in Fig. 4.5 between the ear and cheek are always the leftmost line's endpoint (marked in green). In our case we only have to locate the lowest and highest key-points sets, and select the leftmost points of each set as the edge-points between ear and cheek ($P3$ and $P4$).

4. Locating ROI:

Using the edge points extracted in Step 3, we can segment the ear region and locate a round region on the face. As shown in Fig. 4.6, $P3$ and $P4$ are the leftmost edge points in the lower 1/4 and upper 1/4 of the ear. We connect a straight line from $P3$ to $P4$ as the boundary between ear and face. $P1$ is the lowest edge point and $P2$ is the highest edge point which makes A and B the projection points of $P1$ and $P2$ on the boundary line. C is the lower third of AB, and OC is perpendicular to AB. In our experiment the distance between O and C is 10 mm, and the radius of the circle is 5 mm (as shown in Fig. 4.6). ROI can be constituted by both the ear region and the round region on a face.

5. Ear-Cheek angle calculation:

After preprocessing, we can locate the points in the cheek-plane and detect the boundary points in the ear-plane. To compute the angle the main task is to extract the normal vector that represents the ear/cheek plane. In the ROI selection stage, a set of ear boundary points and a region of cheek points are located, and the 3D coordinates of these points computed using

$$\vec{p} = \begin{bmatrix} x \\ y \\ z \end{bmatrix} = \begin{bmatrix} \frac{x^t(b-d\cdot\tan\theta)}{x^t+f\cdot\tan\theta} \\ \frac{y^t(b-d\cdot\tan\theta)}{x^t+f\cdot\tan\theta} \\ \frac{f^t(b-d\cdot\tan\theta)}{x^t+f\cdot\tan\theta} \end{bmatrix}, \tag{4.2}$$

where (x^t, y^t) are the 2D coordinates of the point on the laser line recorded in the 2D image, b and d are the horizontal and vertical distances between the camera optic center and the projector axes respectively, θ is the laser projection angle, and f is the focal length of the camera. All parameters (b, d, f) are pre-calibrated while θ is controlled by a motor control circuit. Using the 3D coordinates of these points, the cheek-plane and ear-plane can be hypothesized, and the normal vectors of these planes can be computed. PCA has proven to be an effective method in seeking a projection that best represents the data in a least-squares sense by finding the principal: axes of the data matrix. If we treat the 3D coordinates of points in ROI as a 3-by-n matrix, then using PCA on the matrix the first eigenvector (principal component) can be computed as the normal vector which represents the ear/cheek plane. In this chapter we apply the PCA method on the matrix S_e and S_f as shown in Fig. 4.6, and calculate the normal vectors, n_f and n_e. Since n_f is the normal vector of the cheek-plane, and n_e is the normal vector of the ear-plane according to the definition in Sect. 4.2.1, the angle between cheek and ear can be computed. Figure 4.7 is a series of simulations which illustrate the possible angles

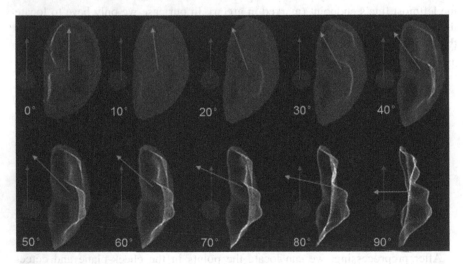

Fig. 4.7 Simulation of various possible ear-cheek angles

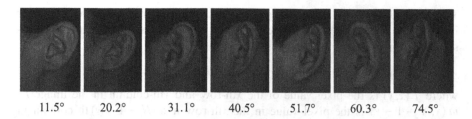

| 11.5° | 20.2° | 31.1° | 40.5° | 51.7° | 60.3° | 74.5° |

Fig. 4.8 Simulation of various possible ear-cheek angles

Fig. 4.9 Same ear at different viewpoints

Table 4.1 Ear-cheek angles of the same ear at different viewpoints

Viewpoints	−20°	−15°	−10°	−5°	0°	5°	10°	15°	20°
Angle (°)	25.61	25.58	24.87	25.60	24.78	24.91	24.46	24.47	24.19
Absolute difference (°)	0.83	0.80	0.09	0.82	0.00	0.13	0.32	0.31	0.59

between ear and cheek. Based on our definition the angle is between 0° and 90°. Some typical ear samples and their respective angles are shown in Fig. 4.8. To test the robustness of the angle calculation, we captured the same person's ear at different viewpoints as shown in Fig. 4.9, and subsequently calculated their ear-cheek angles. The results listed in Table 4.1 show that the angles are robust to viewpoint changes (±20°) while their absolute difference is quite minimal (<0.9°).

4.3 Difference Between Two Ears from the Same Person

In this section the difference between two ears from the same person is illustrated. Section 4.3.1 describes how we normalize a 2D and 3D ear image for matching. Section 4.3.2 then gives matching algorithms for 2D and 3D ears, with the 3D matching algorithm able to better distinguish the same person's two ears.

4.3.1 2D and 3D Ear Image Normalization

To ensure the same scale and viewpoint of 2D and 3D images for each sample, we capture the left and right ears of the same person in a fixed distance and angle.

Then, the 2D and 3D mirror images are calculated from the original right ear images. For the 2D right ear image, I_r, which has rows and columns, its corresponding mirror image, I_m, can be obtained by:

$$f(x,y) = m(x, N-1-y), x = 0 \ldots M-1, y = 0 \ldots N-1, \qquad (4.3)$$

where $f(x,y)$ is the pixel value of the xth-row and yth-column in the image I_r; $m(x, N-1-y)$ is the pixel value in the xth-row and $(N-1-y)$th column in image I_m (as shown in Fig. 4.10).

According to the laser-triangulation imaging principle, the 3D image is reconstructed by 2D image sequences scanned with laser lines. Thus, using the mirror 2D image sequence we can reconstruct the 3D mirror image of the right ear as illustrated in Fig. 4.11. By comparing the mirror models of right ear with left ear we can find out the difference between right and left ears from the same person. Figure 4.12 shows the right ear, left ear and the mirror right ear samples of the same person.

Fig. 4.10 *Right ear* on the *left* and its corresponding mirror image (which becomes the *left ear*) on the *right*

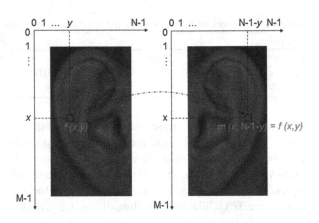

Fig. 4.11 3D mirror ear of the *right ear*

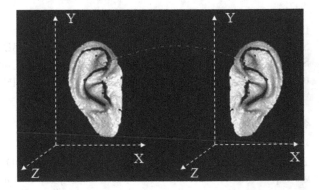

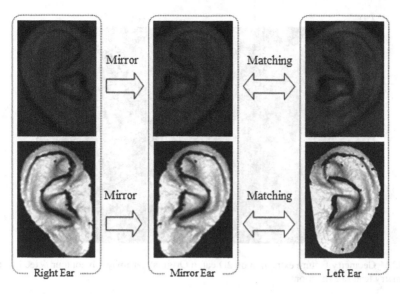

Fig. 4.12 Mirror images of the *right ear*

4.3.2 Distance Between the Two Ears of the Same Person

In 2D ear matching we extract and compare geometry features. Given a 2D ear image (shown in Fig. 4.13a) we wish get a clear ear contour map by first using a Gaussian filter to smooth noise. A canny operator to extract edges from the segmented ear image is next applied (see Fig. 4.13b). Afterwards, a set of points marked in Fig. 4.13c are computed and used to form a feature vector consisting of length, width and area. The Euclidean distance is used to measure the difference between two geometry feature vectors.

For 3D ear matching we use a modified ICP method to measure the distances between two models. The ICP algorithm developed by Besl and Mckay (Besl and McKay 1992) is used to register two given point in a common coordinate system. Each iteration of the algorithm calculates the registration by selecting the nearest points. ICP modified in (Turk and Levoy 1994) is a well-known method to align 3D shapes. Below we explain the algorithm in detail.

1. Initialization of the rotation: matrix $R0$ and translation vector $T0$.
2. Given each point in a test ear, find the corresponding point in the model ear.
3. Discard pairs of points which are too far apart using a tolerance distance *tol*.
4. Find the rigid transformation (R, T) and apply it to the test ear.
5. Go to 2 until no more corresponding points can be found or the maximum iteration number is reached.

In our experiments we used 10 as the tolerance distance for establishing closest point correspondence, and the maximum iteration number is 50 times. After

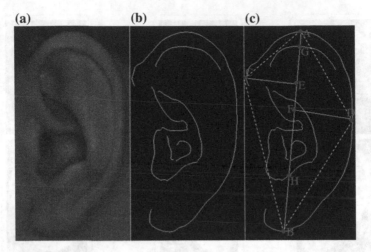

Fig. 4.13 Geometry feature extraction on 2D ear. **a** Original ear image, **b** contour detection, and **c** geometry feature extraction

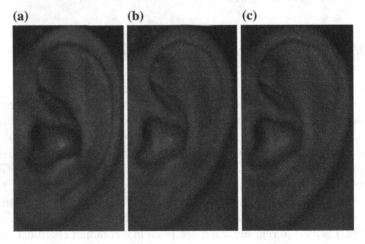

Fig. 4.14 Three samples from *left* and *right ear.* **a** *Left ear*, **b** and **c** *right ear*'s mirror images captured at different times

rotation and translation the average distance of all corresponding points is calculated as the distance of model-test pair.

Figure 4.14 shows three samples from the same person. The first sample (a) is captured from the left ear while the other two samples (b)–(c) are the mirror images from the right ear captured at different times. We used geometry feature distance to measure the difference between 2D ear images, and modified ICP method to measure the distances between their 3D models. The distances between these ears are shown in Table 4.2.

	$D(a,b)$	$D(a,c)$	$D(b,c)$
2D	0.117	0.133	0.130
3D	1.251	1.189	0.667

Table 4.2 2D and 3D distances between left and right ear for the same person

$D(a,b)$, $D(a,c)$ and $D(b,c)$ in Table 4.2 represent the distances between ears in Fig. 4.14a–c. By observing the distances in Table 4.2, we can see that using 2D geometry features the difference between left ear (a) and right ear (b)–(c) cannot be well distinguished. However, with the 3D feature the distance between left and right ear are much larger. Therefore, the 3D ear models are more distinguishable than the 2D images. The 3D ear models not only have geometry features as 2D images but also have depth: information. In the succeeding experiments the modified ICP is used to match all 3D ear models.

4.4 Experimental Results

In the following section we provide the experimental results, which were performed on a single PC with Intel Core 2 CPU at 2.33 GHz and 2 GB of memory. Section 4.4.1 first introduces the ear dataset consisting of 2,000 images. Section 4.4.2 then shows the result of ear-cheek angle for each image. Afterwards, Sect. 4.4.3 provides the matching result of every pair of ears.

4.4.1 3D Ear Dataset

We collected the 3D ear samples from 250 individuals using our 3D ear laser scanner. The subjects mainly consisted of volunteers from students and staff at Shenzhen Graduate School of Harbin Institute of Technology, including 178 males and 72 females with ages ranging from 20 to 60. These samples were collected on two separate occasions, at an interval of around one month. On each occasion the subject was asked to provide two left side face images and two right side face images. Therefore, each person provided 8 images so that our database contains a total of 2,000 images from 500 different ears.

During data collection the subjects sat in a natural position on a chair with a backrest and kept still. The laser scanner was vertically and horizontally movable to accommodate for different seating and head positions. The scanner captures the front ear in a fixed distance of 30 cm and an approximately vertical angle to the side of the face. The subjects were asked to take off all ornaments from their ear and tie their hair back to avoid any occlusions. The scanning process took approximately 2 s.

4.4.2 Ear-Cheek Angle Results

Table 4.3 shows four different samples from the left and right ears of the same person at different capture times. Their ear-cheek angles are also calculated on the second row. Diff_FS represents the angle difference between two ears in the first and the second capture times. Diff_LR means the average angle difference between left and right ears of the same person. We can see that the same ear-cheek angle in different capture times is almost identical. This makes the ear-cheek angle a stable feature. For some people however, their ear-cheek angle of the left and right ears is different. We computed all the sample ear-cheek angles and found the maximum angle deviation for the same ear at different capture times to be $1.72°$. Also, for 81 people their ear-cheek angle was larger than $1.72°$.

We calculated all 500 individuals' left and right ear-cheek angle and plotted its histogram in Fig. 4.15. Table 4.4 shows the ear-cheek angle distribution. The first row shows the range of degrees while the second row depicts the number of individuals that fall into this range. From Fig. 4.15 and Table 4.4 it can be seen that the majority of individuals have an ear-cheek angle between $30°$ and $39°$.

We compared our angle feature with the highest point on the boundary of the ear. The experimental results raised two points: (1) The angle is not correlated to the highest point distance; (2) Indexing with angle is more efficient than the indexing with highest point distance. For (1) we computed all the ear angles and placed them in a set denoted A. At the same time we computed the distance between the highest point and the face plane, and recorded this distance in set D (which has the same ordering as set A). Next, we calculated the correlation coefficient of set A and set D. The correlation coefficient value was 0.073. This proves that the angle metric is extremely uncorrelated to the distance metric, since a correlation coefficient value closer to 1 represents correlation. For (2) we drew the histogram of set D, displayed in Fig. 4.16. From this plot it can be seen that the distance metric

Table 4.3 Ear-cheek angles from the same person

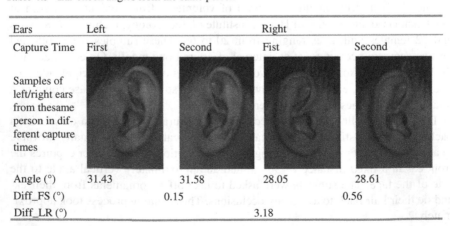

Ears	Left		Right	
Capture Time	First	Second	Fist	Second
Samples of left/right ears from thesame person in different capture times				
Angle (°)	31.43	31.58	28.05	28.61
Diff_FS (°)		0.15		0.56
Diff_LR (°)			3.18	

Fig. 4.15 The angle
histogram distribution plot

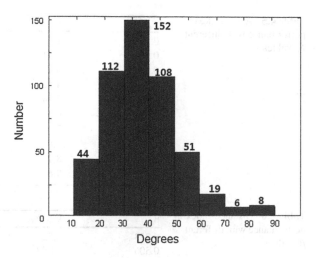

Table 4.4 Ear-cheek angle distribution

Angle (°)	0–9	10–19	20–29	30–39	40–49	50–59	60–69	70–79	80–89	90
Number	0	44	112	152	108	51	19	6	8	0

Fig. 4.16 The distance
histogram distribution plot

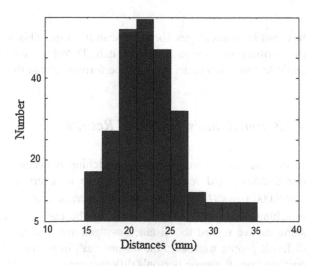

produces a smaller spread than angle. Tables 4.5 and 4.6 are the indexing performances for angle and distance respectively. β_a is the angle search range given
in Fig. 4.2, and is defined as the maximum error value 1.72° (refer to Sect. 4.2)
divided into 10 equal intervals (1/10, 2/10, …, 10/10), while the distance search
range β_d is 2.530. Looking at the middle columns of each table, angle-indexing

Table 4.5 Angle-indexing performance with different β_a values

β_a	k/N (%)	Hit rate (%)
0.172°	1.22	23.6
0.344°	2.41	41.2
0.516°	3.45	61.2
0.688°	4.52	73.6
0.860°	5.44	81.2
1.032°	6.22	86.4
1.204°	7.06	90.0
1.376°	7.88	92.4
1.548°	8.77	93.6
1.720°	9.69	100

Table 4.6 Distance-indexing performance with different β_d values

β_d	k/N (%)	Hit rate (%)
0.253	7.52	24.8
0.506	14.47	46.0
0.759	21.67	61.2
1.012	28.69	72.4
1.265	34.98	82.4
1.518	41.19	88.8
1.771	46.70	91.2
2.024	52.38	94.8
2.277	57.66	97.6
2.530	62.58	100

has smaller percentages for each search range. This means it took less time in the Verification stage to locate a match. Therefore, we can safely assume that the angle feature has a better indexing performance than the distance feature.

4.4.3 Individual's Two Ears' Results

In this experiment we carried out matching between all 2,000 3D ear images in our database, and recorded the distance of every compared pair. This makes 1,999,000 pairs (C_{2000}^2), and with an average matching time of 1.14 s the total matching time took nearly 1 month. If the ear-pair are from the same person's same ear we record the distance as "genuine distance". If the ear-pair are from different people we record the distance as "impostor distance". Finally, if the ear-pair are from the same person's different ears we record the distance as "left–right distance". By analyzing the three categories of ear-pair distances, we can realize the dissimilarity between different ears from same person and from different people. The modified ICP method as described in Sect. 3.2 was used to calculate the distances of the two ears and recorded into genuine, impostor or left–right respectively. This distance distribution curve is shown in Fig. 4.17.

Fig. 4.17 Genuine, *left–right*, and impostor distance distribution *curve*

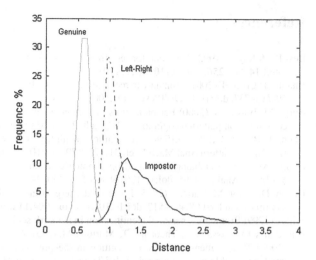

The red curve in Fig. 4.17 is the distance distribution of different samples captured from the same person's same ear or genuine. The minimum distance in this curve is 0.438 while the maximum is 0.811. The green curve represents the distance distribution of various samples captured from the left ear and its corresponding mirror right ear from the same person. The minimum distance here is 0.778 and the maximum distance is 1.473. The Equal Error Rate (EER) between these two curves is 3.37 %, which means the left and right ears are distinguishable. EER is the rate at which both accept and reject errors are equal. The lower the EER value the higher the accuracy of the biometrics system. The blue curve in Fig. 4.17 is the distance distribution of different samples captured from dissimilar people or imposters. The minimum distance in this curve is 0.799 and the maximum distance is 2.879. The EER between genuine and imposter is 2.63 %, signifying an individual's ear can be well differentiated.

4.5 Summary

In this chapter we examined two significant characteristics in a 3D ear: the ear-cheek angle and the dissimilarity between the same person's left and right ear. The ear-cheek angle calculated as the angle between the cheek-plane and ear-plane is unique for each individual, and stable among the same ear captured at different times as shown in the experimental results. This feature is also not available in 2D ear analysis. To compare the same person's left and right ear, a modified ICP method was used, which distinguishes a pair of 3D ears better than a 2D geometry feature extraction method. With an EER of 3.37 %, the left and right ears of the same person can be well separated, while an EER of 2.63 % between genuine and imposter signifies an individual's ear can be well discriminated. Hence, results of the two significant characteristics can be effectively used in 3D ear recognition and indexing.

References

Besl P, McKay N (1992) A method for registration of 3-D shapes. IEEE Trans Pattern Anal Mach Intell 14:239–256. doi:10.1109/34.121791

Bhanu B, Chen H (2007) Human ear recognition in 3D. IEEE Trans Pattern Anal Mach Intell 29:718–737. doi:10.1109/TPAMI.2007.1005

Burge M, Burger W (2000) Ear biometrics in computer vision. In: Proceedings of international conference on pattern recognition, vol 2, pp 822–826. doi: 10.1109/ICPR.2000.906202

Chen H, Bhanu B (2009) Efficient recognition of highly similar 3D objects in range images. IEEE Trans Pattern Anal Mach Intell 31:172–179. doi:10.1109/TPAMI.2008.176

Choras M (2005) Ear biometrics based on geometric feature extraction. Electron Lett Comput Vis Image Anal 5:84–95. doi:10.1007/978-3-540-30074-8_7

Hurley D, Nixon M, Carter J (2005) Force field energy functionals for ear biometrics. Comput Vis Image Underst 98:491–512. doi:10.1016/j.cviu.2004.11.001

Jain A (1999) Biometrics: personal identification in network society. Kluwer, USA

Kakadiaris IA, Passalis G, Theoharis T, Murtuza MN, Lu YL, Karampatziakis N, Toderici G (2007) Three-dimensional face recognition in the presence of facial expressions: an annotated deformable model approach. IEEE Trans Pattern Anal Mach Intell 29:640–649. doi:10.1109/TPAMI.2007.1017

Pang Y, Li X, Yuan Y (2010a) Robust tensor analysis with L1-Norm. IEEE Trans Circuits Syst Video Technol 20:172–178. doi:10.1109/TCSV-T.2009.2020337

Pang Y, Wang L, Yuan Y (2010b) Generalized KPCA by adaptive rules in feature space. Int J Comput Math 87:956–968. doi:10.1080/00207160802044118

Pang Y, Yuan Y, Li X, Pan J (2011) Efficient HOG human detection. Signal Process 91:773–781. doi:10.1016/j.sigpro.2010.08.010

Purkait R, Singh P (2008) A test of individuality of human external ear pattern: its application in the field of personal identification. Forensic Sci Int 178:112–118

Samir C, Srivastava A, Daoudi M (2006) Three-dimensional face recognition using shapes of facial curves. IEEE Trans Pattern Anal Mach Intell 28:1858–1863. doi:10.1109/TPAMI.2006.235

Turk G, Levoy M (1994) Zippered polygon meshes from range images. In: Proceedings of conference on computer graphics and interactive techniques, pp 311–318. doi:10.1145/192161.192241

Yan P, Bowyer KW (2007) Biometric recognition using 3D ear shape. IEEE Trans Pattern Anal Mach Intell 29:1297–1308. doi:10.1109/TPA-MI.2007.1067

Zhang D (2000) Automated biometrics: technologies and systems. Kluwer Academic Publishers, Dordrecht. ISBN 0-7923-7856-3

Zhang D, Kong W, You J, Wong M (2003) Online palmprint identification. IEEE Trans Pattern Anal Mach Intell 25:1041–1050. doi:10.1109/T-PAMI.2003.1227981

Zhang Y, Li Q, You J, Bhattacharya P (2007) Palm vein extraction and matching for personal authentication. Adv Vis Inf Syst 4781:154–164. doi:10.1007/978-3-540-76414-4_16

Zhang D, Lu G, Li W, Zhang L, Luo N (2009) Palmprint recognition using 3-D information. IEEE Trans Syst, Man Cybern Part C: Appl Rev 39:505–519. doi:10.1109/TSMCC.2009.2020790

Chapter 5
3D Ear Feature Extraction and Recognition

Abstract In this chapter, we first define five different features in 3D ear, including point, line and area as local feature, and angle and distance as global feature. Then we discuss the methods to extract these features. The experimental results are given to illustrate the effectiveness of the features. Finally, some applications in indexing and recognition are implemented.

Keywords 3D ear • Feature extraction • Global and local feature

5.1 Introduction

In recent years, some approaches have been developed for ear recognition from 2D images. However, currently avail available range sensors can directly provide us 3D geometric information which is insensitive to above imaging problems. Therefore, it is desirable to extract some useful features and design a human 3D ear recognition system. In fact, different methods to extract features and design biometrics system based on 3D data have been addressed (Huang et al. 2009; Chen and Bhanu 2007; Yan and Bowyer 2007).

Feature extraction is an important stage in 3D ear recognition. If the features extracted are carefully chosen it is expected that the features set will extract the relevant information from the 3D input data in order to perform the desired task using this reduced representation instead of the 3D full size input (Zhang et al. 2009; Besl and Mckay 1992; Bronstein et al. 2003; Chang et al. 2003; Chua et al. 2000). In this chapter, the 3D ear image obtained from Chap. 3 is carefully explored for feature extraction and recognition. Except the characteristics given in Chap. 4 could be as some potential features, a few novel other features could be also illustrated their effectiveness in ear indexing and recognition.

The rest of the chapter is organized as follows. Section 5.2 defines five different features in 3D ear and Sect. 5.3 shows how to extract them, respectively. The

D. Zhang and G. Lu, *3D Biometrics*, DOI: 10.1007/978-1-4614-7400-5_5,   69
© Springer Science+Business Media New York 2013

Table 5.1 Definition of five kinds of 3D ear features

	Global features		Local features		
Types	Angle	Distance	Point	Line	Area
Definition	Angle	Distance from	Key points	5 lines (2 H	Some defined
	between	ear edge	(min/max	and 3 V)	features
	ear-plane	point to	$x/y/z$) with	with all z	in 12
	and cheek	face plate	their z	values on	sub-images
	plane		values	them	
Features	1	1	6	70	96
dimension					

experimental results in Sect. 5.4 demonstrate the effectiveness of the given features. Finally, Sect. 5.4 presents conclusions.

5.2 3D Features Definition

Based on a 3D ear image collected by our laser capturing device, five kinds of features could be defined. The point feature, line feature, and area feature describe key points, shapes and the local area of 3D ears. They will be treated as local features. The angle feature and distance feature represent gesture and scale of a 3D ear, and we treat them as global features. These kinds of features are defined in Table 5.1.

5.3 3D Features Extraction

After the 3D ears are normalized, these defined features could be extracted in the following.

5.3.1 Point Feature

In the normalized *XYZ* coordinate system of a 3D ear we first select 6 points that have min/max x value (points 1 and 2 in Fig. 5.1), min/max y value (points 3 and 4 in Fig. 5.1), and min/max z value (points 5 and 6 in Fig. 5.1). Then, from their z values we can obtain z_1, z_2, z_3, z_4, z_5, and z_6, which combines to form our point feature vector $P(z_1, z_2, z_3, z_4, z_5, z_6)$.

Figure 5.2 shows the point feature vectors extracted from different samples. Sample 1 (S1) and 2 (S2) are from same ear, and Sample 3 (S3) is from a different ear. The red curve is the point feature vector of S1, the blue curve is the point

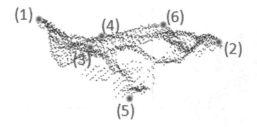

Fig. 5.1   Six key points on the normalized 3D ear

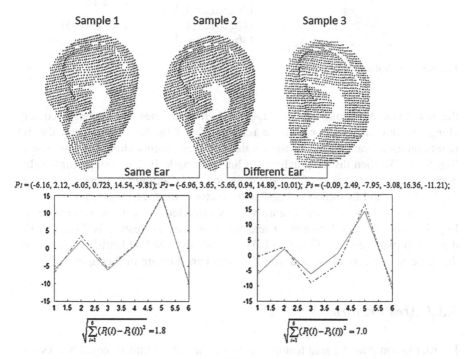

$P_1 = (-6.16, 2.12, -6.05, 0.723, 14.54, -9.81);\ P_2 = (-6.96, 3.65, -5.66, 0.94, 14.89, -10.01);\ P_3 = (-0.09, 2.49, -7.95, -3.08, 16.36, -11.21);$

$$\sqrt{\sum_{i=1}^{6}(P_1(i)-P_2(i))^2} = 1.8 \qquad\qquad \sqrt{\sum_{i=1}^{6}(P_1(i)-P_3(i))^2} = 7.0$$

Fig. 5.2   Point feature vectors extracted from different samples, *Sample* 1 and *Sample* 2 are from same ear, and *Sample* 2 and *Sample* 3 are from different ears

feature vector of S2, and the black curve is the point feature vector of S3. The distance between S1 and S2 is 1.8, and the distance between S1 and S3 is 7.0. We can see that the point feature vectors from the same ear are very similar, and from different ears dissimilar.

5.3.2 Line Feature

To calculate the line feature we first fit a rectangle on the normalized ear in the *XY* coordinates, and define 5 lines, L_1, L_2 (which divides the rectangle equally along

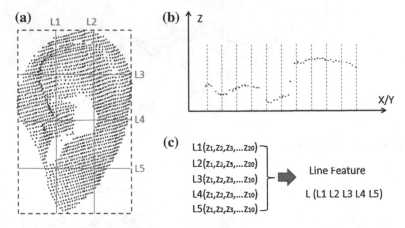

Fig. 5.3 Line feature extraction

the horizontal direction), and L3, L4, L5 (which divides the rectangle equally along the vertical direction). Please refer to Fig. 5.3a. Next, we obtain the 3D points on each line and record the z values of these points along the line (see to Fig. 5.3b). We then divide equally each line and mark the z crossing points value as $z_1, z_2, \ldots, z_{10}$ (or $z_1, z_2, \ldots, z_{20}$ for L_1 and L_2). Combining these z values as L_1, L_2, L_3, L_4, and L_5 forms the line feature vector $L(L_1, L_2, L_3, L_4, L_5)$.

Figure 5.4 shows the line feature vectors extracted from the same samples as Fig. 5.2. The distance between S1 and S2 using the line feature is 7.02, and the distance between S1 and S3 is 41.12. We can see that the line feature vectors from the same ear are very close, and from different ears they are further apart.

5.3.3 Area Feature

In order to compute the area feature we divide the 3D ear into 12 equal blocks (see Fig. 5.5). All points' coordinates in a block as (x_i, y_i, z_i) $i = 1, \ldots, n$, where n is the number of the points in the block. All the coordinates of these points constitute a n-by-3 matrix, M. Then the PCA (Duda 2001; Yang et al. 2004) method is performed on M, and the normal vector $Vn(i, j, k)$ could be represented as follows:

$$M = \begin{bmatrix} x_1 \ y_1 \ z_1 \\ \ldots \\ x_n \ y_n \ z_n \end{bmatrix},$$

then $(\bar{x}, \bar{y}, \bar{z}) = \frac{1}{n} \sum_{i=1}^{n} (x_i, y_i, z_l)$.

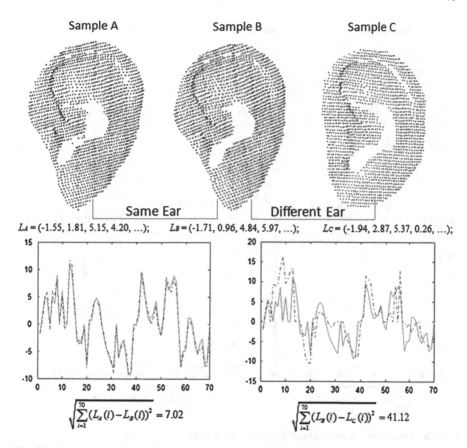

Fig. 5.4 Line feature vectors extracted from S1, S2, and S3

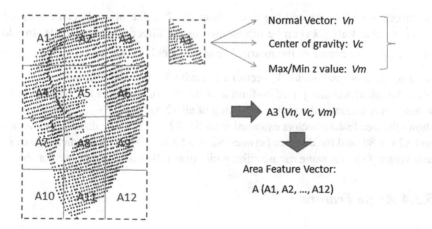

Fig. 5.5 Area feature extraction

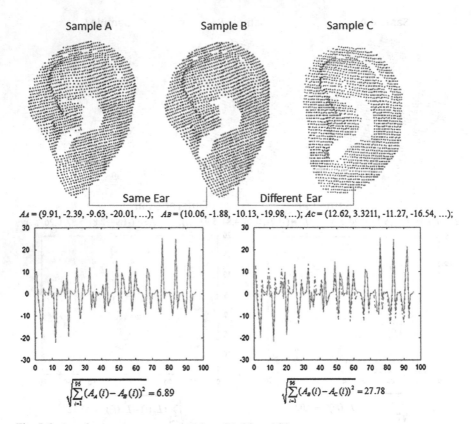

Fig. 5.6   Area feature vectors extracted from S1, S2, and S3

Define the scatter matrix, $S = \sum_{i=1}^{n} (M_i - \bar{M}_i) \times (M_i - \bar{M}_i)^T$; compute the eigenvectors of S as Φ, and the first column of Φ is the normal vector $Vn(i, j, k)$. It is clear that $Vn(i, j, k)$ can be represented as the direction of matrix M. In addition, the gravity center of the matrix M can be calculated as: $Vg = \frac{1}{n} \sum_{i=1}^{n} (x_i, y_i, z_i)$. As a result, a normal vector V_n, center of gravity V_c, and the min/max z values V_m could be calculated and joined to form a vector A_n for each block. The area feature subsequently becomes the vector consisting of all 12 A_n, $A(A_1, A_2, ..., A_{12})$. Figure 5.6 shows the area feature vectors extracted from S1, S2, and S3. The distance between S1 and S2 is 6.89, and the distance between S2 and S3 is 27.78, which indicates area feature vectors from the same ear are alike, while from different ears they are not alike.

5.3.4 Angle Feature

The ear-cheek angle is defined as an angle feature between the normal vector of the cheek-plane and the normal vector of the ear-plane. Details regarding the angle feature can be found in Sect. 4.2.

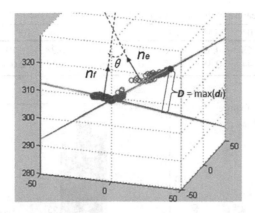

Fig. 5.7 Highest point distance definition

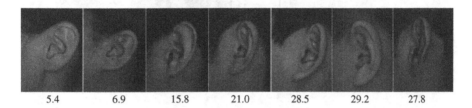

| 5.4 | 6.9 | 15.8 | 21.0 | 28.5 | 29.2 | 27.8 |

Fig. 5.8 Typical ear samples with corresponding distance features

5.3.5 Distance Feature

From the angle feature discussed above we can see that each edge point has a distance to the face plane. We record this distances as d_1, d_2, ..., d_n, and assuming the cheek plane is represented as, $A_f x + B_f y + C_f z + D_f = 0$ the coordinates of edge points becomes (x_i, y_i, z_i), $i = 1,...,n$. Thus, $d_i = \frac{|Ax_i + By_i + Cz_i + D|}{\sqrt{A^2 + B^2 + C^2}}$ $(i = 1, \ldots, n)$, and we define the distance feature value as $D = \max(d_i)$ $(i = 1, \ldots, n)$. Figure 5.7 shows the definition of the distance feature, while Fig. 5.8 illustrates some typical ear samples and their respective distance values.

We will test the robustness of the distance feature, which shows the same person's ear at different viewpoints, and calculate their highest distance. The results listed in Table 5.2 show the maximum difference between the various viewpoints is 0.57 mm.

We calculated everyone's highest distance feature and plotted the results in a histogram shown in Fig. 5.9. The numeric breakdown of this plot is depicted in Table 5.3, where the first row represents the distance and the second its number.

Table 5.2   Distance feature of the same ear at different viewpoints

Viewpoints	−20°	−15°	−10°	−5°	0°	5°	10°	15°	20°
Distance (mm)	10.58	10.62	10.56	10.71	10.14	10.28	10.19	10.07	9.92
Difference (mm)	0.44	0.48	0.42	0.57	0.00	0.14	0.05	−0.07	−0.22

Fig. 5.9   The distance feature histogram distribution

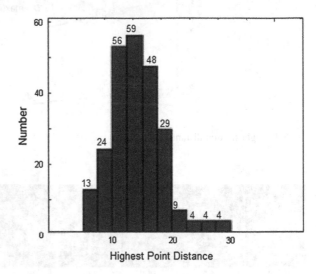

Table 5.3   Distance feature distribution

Distance (mm)	0	5	7.5	10	12.5	15	17.5	20	22.5	25	27.5	30
Number	0	13	24	56	59	48	29	9	4	4	4	0

5.4 Experimental Results

The experiments were divided into two parts: recognition experiments using global feature indexing, and verification experiments with local feature matching. In addition, the fusion of global and local features was performed and presented in the verification part.

Our database contains a total of 500 samples from 250 different ears. We used $E1$ as the test ear dataset and $E2$ as the model ear dataset. $E1$ contains 250 samples from 250 different ears, while $E2$ contains 250 samples from the same 250 ears as $E1$, but captured at different times from $E1$. A PC with Intel Core 2 CPU @2.33 GHz and 2 GB memory was used in our experiments.

5.4.1 Global Feature Indexing

According to the histogram of both angle and distance features, the ear-cheek angle ranges from 0 to 90° for 3D ears with its highest point distance from 0 to

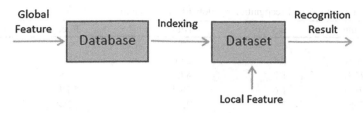

Fig. 5.10   Flowchart of the recognition with global feature indexing

30 mm. These global features are stable for the same person and distinguishable for different individuals, therefore, the angle and distance values can be initially used for 3D ear recognition. The flowchart of this recognition with global feature indexing is shown in Fig. 5.10. For a given test ear sample:

1. Extract the global features of the test sample: $G_t(A, D)$;
2. Compare the global features $G_t(A_t, D_t)$ with all model ears' global features $G_i(A_i, D_i)$ $i = 1,...,N$. (In our experiments $N = 250$. Distance is measured as a vector $Dist(|A_i- A_t|, |D_i- D_t|)$;
3. If $Dist(|A_i- A_t|, |D_i- D_t|)$ is smaller than the threshold $T(\beta, \gamma)$, we treat the model ear as a matched candidate and pace it into a sub-database;
4. Extract the local features of test ear V_t and the local features of the model ears in the candidate sub-database V_i $(i = 1,...,k)$, k is the total number of ears in candidate sub-database;
5. Perform local feature matching between V_t and V_i to measure the differences between test ear and candidate ears. (In our experiments the Euclidean distance was used to measure the differences between V_t and V_i);
6. The candidate ear with the smallest difference to the test ear is assigned as the recognition result.

In our experiments, 10 different thresholds $Ti(\beta, \gamma)$ $(i = 1,...,10)$ are used to evaluate the indexing performance. For a smaller threshold, the matching time in recognition is quicker, while the larger the threshold is the higher the recognition rate achieved. Because when the threshold is smaller, the number of candidate ears would be less. So the smaller threshold leads to less Local Feature extraction and matching times in recognition. On the other hand, the small threshold would reject the genuine samples incorrectly, and affect the recognition rate.

We computed all the global feature differences between $E_1(i)$ and $E_2(i)$ $(i = 1,...,250)$, and found the maximum angle difference between the same ears at different capture times to be $1.72°$, and the maximum distance difference to be 2.53 mm. Next, we divided the maximum angle value and distance value into 10 equal intervals, and set $T_i(\beta, \gamma)$ as $i \times (1.72, 2.53)$ $(i = 1,...,10)$. Given a test sample in E_1, based on different $T_i(\beta, \gamma)$ values we can obtain a sub-database which contains k candidates among the total number N (in our database $N = 250$) in E_2. If the corresponding object (model sample captured from the same ear as the test ear) is in the list of the k candidates, we take the indexing result as being correct.

Table 5.4 Performance of recognition by global feature indexing

$T(\beta, \gamma)$ (°, mm)	k/N (%)	Hit rate (%)	Recognition rate (%)
(0.172, 0.253)	0.1	7.2	7.2
(0.344, 0.506)	0.4	23.6	23.6
(0.516, 0.759)	1.2	44.8	44.8
(0.688, 1.012)	1.6	58.4	58.4
(0.860, 1.265)	2.4	72.4	72.4
(1.032, 1.518)	2.8	77.2	76.0
(1.204, 1.771)	3.2	82.4	81.2
(1.376, 2.024)	4.4	88.4	87.2
(1.548, 2.277)	5.2	92.4	91.2
(1.720, 2.530)	6.4	100	98.4

The indexing performance is evaluated by computing the average ratio between the number of correctly indexed objects and the total number of test objects. We record the average ratio as the hit rate.

For example, given a test sample with global feature $G_t = (35.38, 26.18)$, and a threshold $T_i(\beta, \gamma) = (1.72, 2.53)$, the candidate ears are models which have angle values between 35.38 ± 1.72 and distance values between 26.18 ± 2.53 mm.

Table 5.4 shows the indexing and recognition performances using different thresholds. The first column shows 10 different thresholds $T_i(\beta, \gamma)$, second column illustrates the average matching times compared to the total matching times, third column is the hit rate which presents the correct indexing ratio, and the last column depicts the recognition rate under different indexing thresholds.

5.4.2 Matching by Local and Global Features

The following experiments were performed using local (point, line, area), and global features (angle, distance), as well as their feature-level and score-level fusion.

First, the point, line, and area features are used separately as the feature vector of a 3D ear, by matching each ear with all other ears in our database. In matching stage, we used Euclidean distance to measure different feature vectors. Afterwards, we merged all the local features into one feature vector, and used this fusion vector as the local feature. From our definition, the point feature vector has 6 components, the line feature vector has 70 components, and the area feature vector has 96 components. Therefore, the local feature is described as a 172 long vector. In the fusion feature matching stage, we also computed the Euclidean distance to measure distance.

Since the global features only have 2 components, if we merged the global feature vector into a local feature vector, it would play a small role in the matching stage. Hence, we used the score level fusion instead of feature fusion

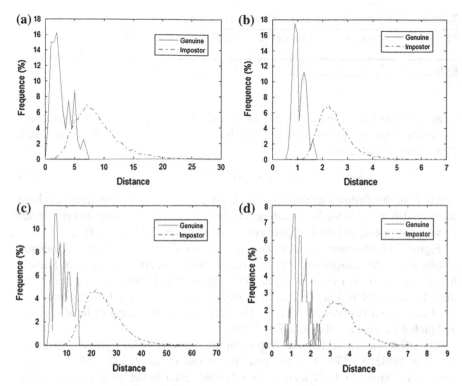

Fig. 5.11 Genuine-Impostor distribution curves by local features: **a** With point feature, **b** With line feature, **c** With area feature, and **d** Fusion of local features (point + line + area)

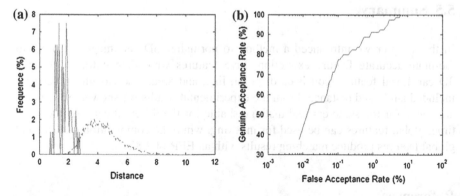

Fig. 5.12 Fusion of local and global features: **a** Genuine-Impostor distribution, and **b** ROC *curve*

Table 5.5 EER values by using different feature extraction methods

Features	Point (%)	Line (%)	Area (%)	Local (%)	Local + global (%)
EER	12.59	3.54	6.31	3.16	2.82

to achieve the fusion matching of global and local features. In our experiment we applied the weighted plus method for score level fusion (Ross and Jain 2003; Zhang et al. 2010):

$$Score\_F = \lambda \times Score\_G + (1 - \lambda) \times Score\_L, \tag{5.1}$$

Score_F is the fusion matching score, *Score_G* is the global feature matching score, and *Score_L* is the local feature matching score. λ is the weighted parameter in our experiment, and the best performance was achieved when λ is 0.3.

Figure 5.11 illustrates the genuine and impostor distribution of the verification results in 3D ear recognition. Figure 5.11a–c show the curves using point feature, line feature, and area feature respectively. Figure 5.11d illustrates the curves by their feature-level fusion. Figure 5.12a depicts the score-level fusion combining local and global features of 3D ears. The ROC curves by fusion of local features and global features are shown in Fig. 5.12b.

The EER values which are important to the verification performance are listed in Table 5.5. We can see that the fusion of local and global features achieve the smallest EER among all schemes, even better than single feature matching. This is reasonable because more information usually leads to a more accurate recognition.

5.5 Summary

In this chapter we introduced a method to normalize 3D ear images in order to facilitate accurate feature extraction. Five features were then extracted from a 3D ear. Local features consisted of point, line, and area, whereas global features included angle and distance. From the experimental results we showed all features are stable for the same ear and distinguishable for the different ears. At the same time, global features can be used for indexing, while the combination of local and global features produce matching results with an EER of 2.82 %.

References

Besl P, Mckay ND (1992) A method for registration of 3-D shapes. IEEE Trans Pattern Anal Mach Intell 14(2):239–256. doi:10.1109/34.121791

Bronstein A, Bronstein M, Kimmel R (2003) Expression in variant 3D face recognition. Audio and video based biometric person authentication 62–70. doi: 10.1007/3-540-44887-X_8

Chang KC, Bowyer KW, Sarkar S, Victor B (2003) Comparison and combination of ear and face images in appearance-based biometrics. IEEE Trans Pattern Anal Mach Intell 25(9):1160–1165. doi:10.1109/TPAMI.2003.1227990

Chen H, Bhanu B (2007) Human ear recognition in 3D. IEEE Trans Pattern Anal Mach Intell 29(4):718–737. doi:10.1109/TPAMI.2007.1005

Chua CS, Han F, Ho Y (2000) 3D human face recognition using point signatures. In: Proceedings of international conference on automatic face and gesture recognition, pp 233–238. doi:10.1109/AFGR.2000.840640

Duda RO (2001) Pattern classification, 2nd edn. Wiley, ISBN: 0471056693

Huang C, Lu G, Liu Y (2009) Coordinate direction normalization using point cloud projection density for 3D ear. In: Proceedings of international conference on computer sciences and convergence information technology, pp 511–515. doi: 10.1109/ICCIT.2009.56

Ross A, Jain A (2003) Information fusion in biometrics. Pattern Recogn Lett 24(13):2115–2125. doi:10.1016/S0167-8655(03)00079-5

Yan P, Bowyer KW (2007) Biometric recognition using 3D ear shape. IEEE Trans Pattern Anal Mach Intell 29(8):1297–1308. doi:10.1109/TP-AMI.2007.1067

Yang JY, Zhang D, Frangi AF, Yang J (2004) Two-dimensional PCA: a new approach to appearance-based face representation and recognition. IEEE Trans Pattern Anal Mach Intell 26(1):131–137. doi:10.1109/TPAMI.2004.1261097

Zhang D, Lu G, Li W, Zhang L, Luo N (2009) Palmprint recognition using 3-D information. IEEE Trans Syst Man Cybern C: Appl Rev 39(5):505–519. doi:10.1109/TSMCC.2009.2020790

Zhang D, Guo Z, Lu G, Zhang L, Zuo W (2010) An Online System of Multispectral Palmprint Verification. IEEE Trans Instrum Meas 59(2):480–490. doi:10.1109/TIM.2009.2028772

Part III
3D Palmprint Authentication Using Modulated Structured Light

Chapter 6
3D Palmprint Capturing System

Abstract Palmprints have been widely studied for personal authentication because they are highly accurate and incur low costs. Most of the previous work has focused on two-dimensional palmprint identification. However, the inner surfaces of palms not contain only texture information, but also shape information. Unfortunately, two-dimensional palmprint systems lose the shape information when capturing palmprint images. Hence, three-dimensional information is important for palmprint systems. In this chapter, we have designed and developed a novel three-dimensional palmprint acquisition system based on structured-light imaging technology. The acquisition system can obtain palmprint three-dimensional information and at the same time, the corresponding two-dimensional texture, which are used for personal authentication. A three-dimensional palmprint database has been established by using the developed acquisition system, and the testing results illustrate the effectiveness of our system.

Keywords 3D palmprint measurement • Biometrics • Structured-light imaging • Palmprint depth

6.1 Introduction

As a physiological biometrics characteristic, the palmprint was proposed for personal recognition more than 10 years ago (Shu et al. 1998), and has been widely studied due to its merits, such as distinctiveness, cost-effectiveness, user friendliness, high accuracy, and so on.

Most of the previous work has focused on two-dimensional (2D) palmprint identification. Kong et al. (2004) proposed a competitive coding scheme for palmprint verification. Sun et al. (2005) proposed an ordinal palmprint representation for personal identification. Jia et al. (2008) suggested a robust

D. Zhang and G. Lu, *3D Biometrics*, DOI: 10.1007/978-1-4614-7400-5_6,
© Springer Science+Business Media New York 2013

line orientation code method   for palmprint verification. There are mainly
three ways to obtain 2D palmprint images with the following advantages and
disadvantages:

1. Ink-based images (Shu et al. 1998; Zhang et al. 1999). Advantages: high resolution
   and discriminability. Disadvantages: low user acceptability and obtaining speed.
2. Scanner-based images (Fratric et al. 2008). Advantage: low cost. Disadvantage:
   low obtaining speed.
3. Charge-coupled device (CCD) camera-based images (Fratric et al. 2008; Zhang
   et al. 2003). Advantages: high obtaining speed and image quality. Disadvantage:
   none.

Figure 6.1 shows a CCD camera-based 2D palmprint acquisition device (left)
developed by the Hong Kong Polytechnic University and a palmprint image (right)
collected by this device (Wong et al. 2005). An online palmprint identification system was then developed based on this CCD camera-based 2D palmprint acquisition device (Zhang et al. 2003).

Although 2D palmprint recognition has proven to be efficient in terms of verification rate, it has some inherent drawbacks. First, the palm is not a pure plane and
three-dimensional (3D) depth information cannot be captured by using a single
CCD camera. Secondly, the illumination variations in the system will substantially
affect the 2D palmprint image and may lead to false recognition. Then, although
the area of palm is large, too much contamination or scrabbling in the palm can
still render the recognition invalid. Lastly, the 2D palmprint image can be easily
copied and counterfeited; hence, the anti-forgery abilities of 2D palmprint devices
need to be improved.

From inked palmprint samples, we can see that the inner surface of the palm
contains consistent and detailed shape information Fig. 6.2 shows some inked

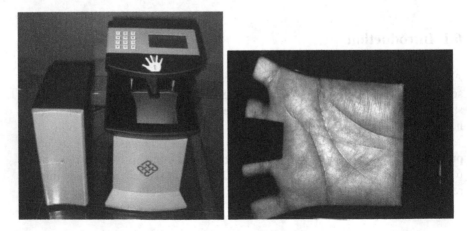

Fig. 6.1 CCD camera-based 2D palmprint acquisition device (*left*) and a palmprint image
(*right*) that is collected by the device

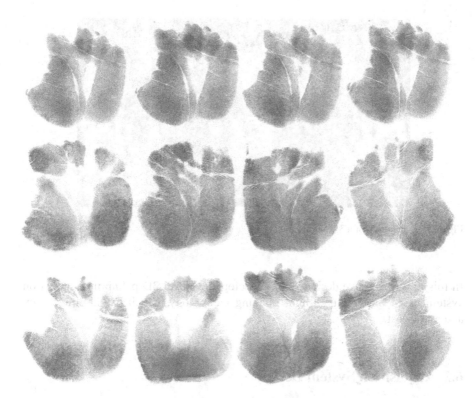

Fig. 6.2 Inked palmprint samples, the *first row* shows prints that are collected from the same palm at different times, and the *second* and *third rows* show prints that are collected from different palms

palmprint samples, in which the prints in the first row are collected from the same palm at different times, and the prints in the second and third rows are collected from different palms. From these samples, we find that there are some blank areas in the palm centers which are consistent for the samples collected from the same palm, and vary for the samples collected from different palms. This implies that the depth of the inner surface of the palm is useful information for personal authentication.

Figure 6.3 depicts a 3D palmprint surface from which we can see that the inner surface of the palm is really not a plane, but contains a substantial amount of shape information. There are some commercial devices which can obtain the 3D information of an object, such as Konica Minolta Vivid 9i/910 (Konica Minolta Sensing 2010), Cyberware whole body color 3D scanner (Cyberware Inc. 2010), and so on. These commercial 3D scanners have high speed and accuracy. However, they are very expensive and cumbersome. As they are not specifically designed for collecting palmprints, they cannot automatically detect whether the palm is placed in the correct position. Hence, it is not feasible to establish an automatic palmprint authentication system by using commercial 3D scanners.

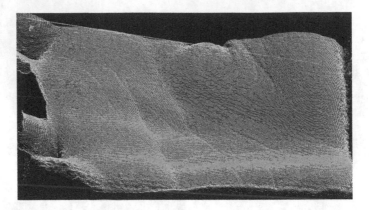

Fig. 6.3   A depiction of a 3D palmprint surface

In this paper, we have designed and developed a novel 3D palmprint acquisition system based on structured-light imaging technology, which has high accuracy and a low cost.

6.2   Acquisition System Design

6.2.1   Feasibility Analysis

Non-contact vision-based 3D imaging technology primarily incorporates multi-view imaging and structured-light imaging. Multi-view imaging is based on human visuals. It is high in speed and low in cost. However, it is hard to obtain high accuracy because it is difficult to find and match corresponding point pairs in two or more images. Structured-light imaging is widely used as a 3D imaging method for its high accuracy, speed and stability. A projector casts a certain pattern of structured light onto a surface, the structured light is modulated by the surface shape, and the modulated stripes are captured by a CCD camera at a constant distance from the projector. The distance from the measured surface to the reference plane can be calculated according to the modulated stripe images and the geometric correlation between the measured surface, projector and CCD camera.

There are four kinds of 3D imaging models according to the structured light patterns: point structured light, line structured light, multiple line structured light and grid structured light, as shown in Fig. 6.4 Point structured light is the simplest model. It is very easy to calculate the point position according to the triangulation. However, it has low efficiency as it only measures one point each time. Line structured light is the extension of point structured light. We must let the line light

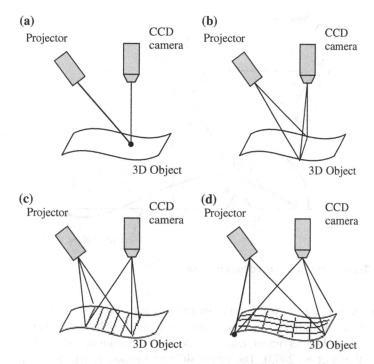

Fig. 6.4 Four types of structured-light imaging models. **a** Point structured light. **b** Line structured light. **c** Multi-line structured light. **d** Grid structured light

scan the measured surface and record the images by a high speed CCD camera to obtain the 3D information of objects. So, it is still not very efficient. Multi-line structured light can obtain all of the 3D information by one or several images, which is very efficient and accurate. Although grid structured light is also efficient, it is much too complex for application. So, we choose multi-line structured light to establish the system.

Figure 6.5 illustrates the imaging principle of the multi-line structured-light technique (Srinivassan et al. 1984). Interested readers can refer to Srinivassan et al. (1984) for more details. In Fig. 6.5, there is a reference plane in which the height is 0. By projecting light through the grating to the object surface, the relative height of point D at spatial position (x, y) to the reference plane can be calculated as follows (Srinivassan et al. 1984)

$$h(x, y) = \overline{BD} = \frac{P_0 \cdot \tan\theta_0 \cdot \phi_{CD}}{2\pi(1 + \tan\theta_0 / \tan\theta_n)}. \tag{6.1}$$

where P_0 is the wavelength of the projected light on the reference plane, θ_0 is the projecting angle, θ_n is the angle between the reference plane and the line which passes through the current point and the CCD center, and ϕ_{CD} is the phase difference between points C and D.

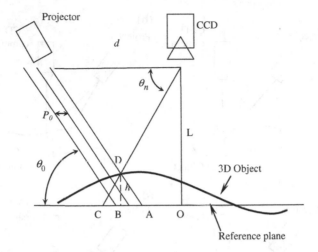

Fig. 6.5 The principle of structured-light imaging

From Eq. 6.1, we can see that the key to this method is to solve the phase information. There are mainly three methods to solve the phase information of the grating image: Fourier transform, time domain convolution filtering and phase shifting (Fiona et al. 2003). The phase shifting method is widely recognized as the most effective and reliable method (Hung et al. 2000; Judge et al. 1994). It has low computational complexity and capacity of anti-static noise. Hence, we adopt the phase shifting method to calculate the phase information (Huntley et al. 1993). With a four-step interferometer (Huntley et al. 1993), four intensity images, I_1, I_2, I_3, I_4, are obtained by projecting four sinusoidal grating patterns onto the object, separated by the phase steps of $\pi/2$:

$$I_n(x,y) = I_r + I_o + 2A_rA_o\cos\left(\phi(x,y) + \delta_n\right), \quad \delta_n = (n-1)\pi/2,$$
$$n = 1,2,3,4, \tag{6.2}$$

where A_r and A_o are the reference and object beam amplitudes with corresponding intensities I_r and I_o. I_n, I_r, I_o, A_r, A_o and ϕ are all functions of x and y. According to Eq. 6.2, the phase mapping can be calculated as:

$$\phi(x,y) = \tan^{-1}\left(\frac{I_4(x,y) - I_2(x,y)}{I_1(x,y) - I_3(x,y)}\right). \tag{6.3}$$

From Eq. 6.3, we can see that all of the phase values fall into the interval of $[-\pi/2, \pi/2]$. In order to obtain a continuous distribution of phase values, the phase values first need to be extended to $[-\pi, \pi]$. This can be done according to the signs of the sine and cosine values of the phase values. Then, we use the phase unwrapping method (Tribolet 1977) to retrieve the continuous phase mapping. The main steps are as follows.

Figure 6.6 illustrates the principle of phase unwrapping where Fig. 6.6a shows

Fig. 6.6 The principle of phase unwrapping

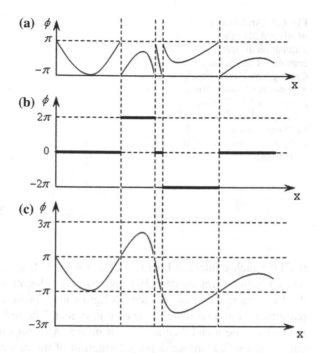

an example of calculated phase values which are not continuous. We can see that there must be a step of 2π for very discontinuous points, as shown in Fig. 6.6b. If we can correctly compensate the step at the discontinuous points, we can obtain the original continuous phase values, as shown in Fig. 6.6c. The process of the phase unwrapping method is as follows:

Step 1. Set threshold T at a little less than 2π, e.g. $T = 1.8\pi$;
Step 2. Start from $i = 0$, do the following Steps 3, 4 and 5 until all the phase values have been accessed;
Step 3. Calculate the phase difference of the adjacent points:

$$\Delta\phi = \phi(x_{i+1}, y) - \phi(x_i, y). \tag{6.4}$$

Step 4. If $|\Delta\phi| < T$, then $\phi(x_{i+1}, y) = \phi(x_{i+1}, y)$, else.

$$\phi(x_{i+1}, y) = \begin{cases} \phi(x_{i+1}, y) - 2\pi, & \text{if } \Delta\phi > 0 \\ \phi(x_{i+1}, y) + 2\pi, & \text{if } \Delta\phi < 0 \end{cases} \tag{6.5}$$

Step 5. $i = i + 1$, go to Step 3.

6.2.2 Components of the System

Figure 6.7 illustrates the architecture of the developed 3D palmprint data acquisition system. The system consists of two parts, the projecting and image capturing units. The projecting unit contains the casting lens, liquid crystal display

Fig. 6.7 Architecture of the developed 3D palmprint data acquisition device: *1* CCD camera, *2* Camera lens, *3* Casting lens, *4* LCD panel, *5* Controller board, *6* Back convergent lens, *7* Front convergent lens, *8* White LED light source, *9* Signal and power control box, and *10* Box shell

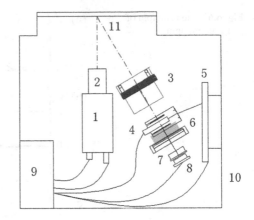

(LCD) panel, controller board, back convergent lens, front convergent lens and white light-emitting diode (LED) light source. The image capturing unit contains the CCD camera and camera lens. A light source projects varying structured light patterns (stripes) onto the surface of an object. The reflected light is captured by the CCD camera and then a series of images are collected. After performing calculations, the 3D surface depth information of the object can be obtained. In earlier stages, parallel lights, such as lasers or point light arrays were used. With the development of light source techniques, liquid crystal light projectors have been successfully used as the new light source (Sansoni et al. 1994). In our developed system, a cost-effective grey LCD projector with an LED light source is employed, and some shift light patterns are projected onto the palm.

Figure 6.8 shows the main components of the controller board. An RGB sinusoidal grating signal is sent to the sample/hold circuits after gamma correcting. Then, the clock generator controls the sample/hold circuits to send an analog image to the LCD. Finally, the projector casts the image on the LCD to the measured object surface. The clock generator determines the time sequence by horizontal synchronization (HSYNC) and vertical synchronization (VSYNC). Figure 6.9 shows the controller board and the LCD.

6.2.3 System Calibration and Measurement

Calibration is an important step for 3D object measurement. Here, we use a comprehensive calibration method. There are three parameters, P_0, d and L, which need to be calibrated as shown in Fig. 6.5, where P_0 is the wavelength of the projected light on the reference plane, d is the distance from the center of the CCD camera to the center of the projector, and L is the distance from the center of the CCD camera to the reference plane. According to triangular correlation, we can get the following equation:

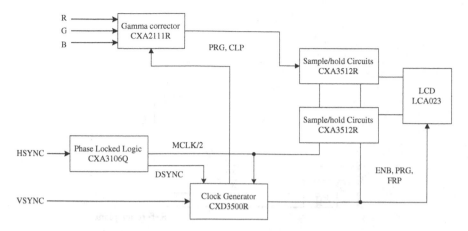

Fig. 6.8 Main components of the controller board

Fig. 6.9 The controller board and the LCD

$$\frac{L - h}{h} = \frac{d}{\frac{\phi_{CD}}{2\pi} \cdot P_0},$$ (6.6)

that is

$$h = \frac{\phi_{CD} \cdot L \cdot P_0}{2\pi d + \phi_{CD} \cdot P_0}$$ (6.7)

From Eq. 6.7, we can see that if h and ϕ_{CD} are known, the unknown parameters are P_0, d and L. Actually, if a regular target is measured, its height is known, and ϕ_{CD} can be calculated by the collected images. So, given three different heights, h_1, h_2, h_3, as shown in Fig. 6.10, we can get three equations from Eq. 6.7, then, the three unknown parameters, P_0, d and L, can be resolved.

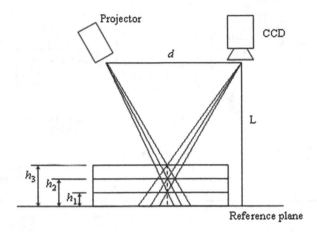

Fig. 6.10 Illustration of the *three heights* for calibration

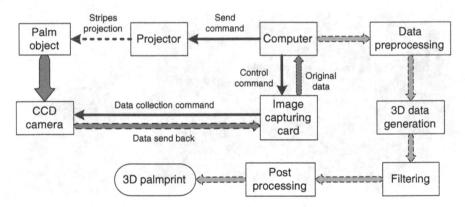

Fig. 6.11 3D palmprint data collection and processing process (the *red solid line arrows* denote sending command, *green arrows* denote the data transport of collecting, and *blue arrows* denote the data processing)

Figure 6.11 illustrates the 3D palmprint data collection and processing process. The computer controls the projector to cast a series of 13 structured-light stripes onto the palm inner surface and the CCD camera captures the palm images with the projected stripes. At the same time, the computer sends a command to the data collection board to store the images. The data collection requires about 2 s. From these palm images, the depth information of each point on the palm can be computed using phase transition and phase expansion techniques (Huntley et al. 1993; Tribolet 1977). The processing steps, which are marked using blue arrows in Fig. 6.11, will require about 0.5 s. Hence, the total time to generate a 3D palmprint is about 2.5 s.

Figure 2.6 shows the developed 3D palmprint authentication system. A series of the 13 palmprint images with different stripes on them which are used to generate 3D palmprints are given in Fig. 6.12. Figure 6.13 indicates some 3D palmprint samples and their corresponding 2D images.

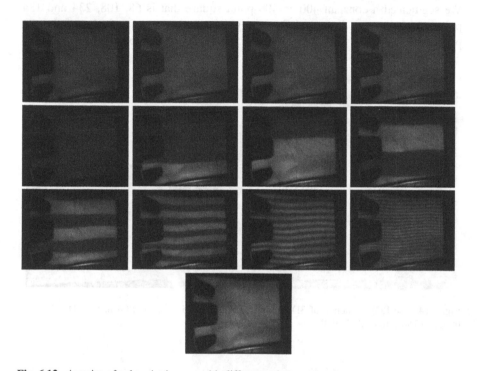

Fig. 6.12   A series of palmprint images with different stripes on them

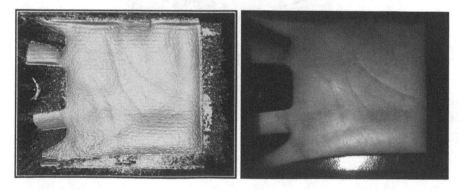

Fig. 6.13   3D palmprint samples and their corresponding 2D images

6.3 Parameter Selection

The original 3D palmprint collected by the developed device contained
768 × 576 points. First, we removed the redundant and noisy boundary regions
using a very simple region of interest (ROI) extraction process (Fig. 6.14).
We segmented a constant 400 × 400 point square that is 68, 108, 234 and 134
points from the top, bottom, left and right boundaries, respectively, of the 3D
palmprint image as shown in Fig. 6.15a, b shows the extracted ROI. After

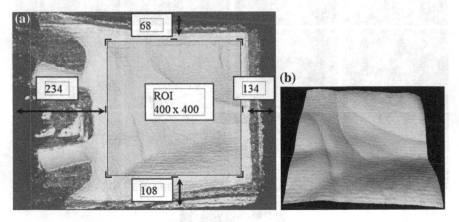

Fig. 6.14 The ROI extraction of 3D palmprint. **a** The location of the ROI in the 3D palmprint
image, **b** The extracted 3D ROI

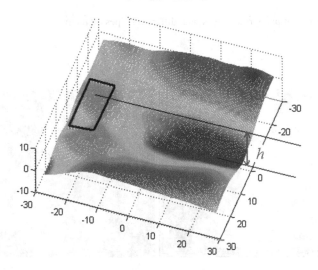

Fig. 6.15 Illustration of the maximum depth value

downsampling the 3D ROI to 200 × 200 points, we store it in a 200 by 200 matrix, $\{d_{ij} \mid i = 1,2,\cdots,200;\ j = 1,2,\cdots,200.\}$, where d_{ij} is the depth value of the ith row and jth column point of the 3D ROI.

With the obtained 3D ROI, we calculated the maximum depth value of the 3D palm from a reference plane, as shown in Fig. 6.15. We tested the depth of 100 palms. From each palm, 10 samples were collected. Figure 6.16 shows the statistical result, from which we can see that the depth value is consistent for the same palm while varying for different palms.

To describe the shape of the 3D palmprint, we used a group of equidistant horizontal planes to cut the 3D ROI as shown in Fig. 6.17. Then, Fig. 6.19 gives the contours that are cut by the equidistant horizontal planes which are collected at different times. The top six contours are obtained from one palm, and the bottom

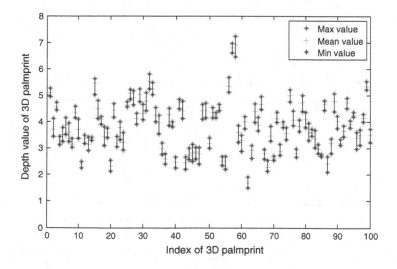

Fig. 6.16 Illustration of the maximum value, mean value and minimum value of the 3D palmprint depths of 100 palms, with 10 3D palmprint samples for each palm

Fig. 6.17 Illustration of the 3D ROI crossed by *horizontal plane*

six contours are obtained from another palm. From Fig. 6.18, we can see that the contours are consistent for the same palm and vary for the different palm.

If we normalize the contours to 32 grades, then we get 32 isodepths, which imply that we give every point a grade value, as shown in Fig. 6.19. The 32 isodepths in the database are denoted by C_d and the 32 isodepths of the input testing sample are denoted by C. The matching score between C_d and C_t can be defined as:

$$R = \frac{\sum\limits_{i=1}^{n} \sum\limits_{j=1}^{m} |C_d - C_t|}{L \times n \times m}, \tag{6.8}$$

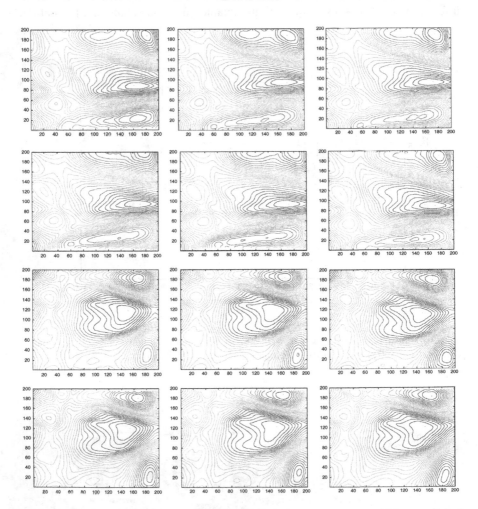

Fig. 6.18 Contours of 3D ROIs that are cut by equidistant horizontal planes which are collected at different times (the *top six* contours are obtained from one palm, the *bottom six* contours are obtained from another palm)

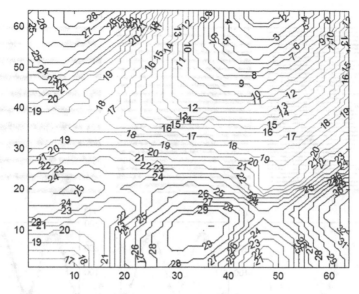

Fig. 6.19   Illustration of 32 isodepths

where L is the level number and $n \times m$ is the size of the matched samples. Through this matching method, the equal error rate (EER) is 3.36 % on a database which contains 1,000 samples collected from 100 palms.

6.4   System Performance Analysis

6.4.1   Grating Wavelength

From Eq. 6.1, we can see that the phase value is the most important factor for calculating the depth information of the measurement object. Theoretically speaking, measurement accuracy is independent from the grating period. As the image obtained by the CCD camera is not continuous but discrete, however, the grating period needs to be taken into consideration.

As described in Sect. 6.2.1, sinusoidal grating is expected. We separately tested the wavelengths of the grating for 8, 16, and 32 pixels as shown in Fig. 6.20. Wavelengths of 2 and 4 imply square and triangular waves which greatly deviate from the sinusoidal wave. Figure 6.20a–c show gratings that are cast by a projector whose wavelengths are 8, 16, and 32 pixels, respectively. Two periods of wave curves along the vertical direction are depicted in Fig. 6.20d–f for the three grating wavelengths, respectively. Figure 6.20g–l indicate the corresponding grating images and two periods of wave curves as captured by the CCD camera. From Fig. 6.20, we can see that for the grating projected by the projector, the accuracy

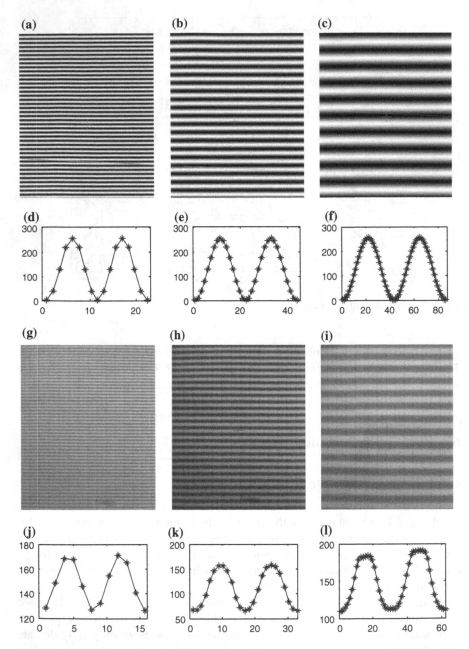

Fig. 6.20 Analysis of different wavelengths of grating

of the sinusoidal curve is improved with the growth of the wavelength. However, as limited by the electrical devices, the accuracy of the captured image is not always improved with the growth of the wavelength, as shown in Fig. 6.20g–i.

Table 6.1 Root mean square deviation of different wavelengths

Wavelength of the grating (pixels)	8	16	32
Wavelength of the projected light on the reference plane (mm)	0.9	1.7	3.4
Root mean square deviation	2.22	0.96	5.28

Table 6.1 lists the root mean square deviation of different wavelengths for captured images by the CCD camera. From Table 6.1, we can see that the root mean square deviation is minimized when the wavelength of the grating is 16 pixels and the wavelength of the projected light on the reference plane is 1.7 mm. So, in our system, we set the wavelength of the grating to 16 pixels.

6.4.2 Signal to Noise Ratio

Another very important factor for system accuracy is the signal to noise ratio (SNR) of the CCD camera. We tested four CCD cameras which had SNRs of 54, 56, 58 and 60 db, respectively. A smooth plane was measured by the system with the different cameras. Figure 6.21 shows the curve of the standard deviation to the SNR. We can see that with the growth of the SNR, the standard deviation of the data decreases, which means that accuracy is improved. However, the cost of a high SNR CCD camera is expensive. In consideration of the performance and cost, we chose the 58 db SNR CCD camera. The depth precision of the 3D image measured by this system is between 0.05 and 0.1 mm.

6.4.3 Data Analysis

A 3D palmprint database has been established by using the developed 3D palmprint imaging device. The database contains 8,000 samples from 200 volunteers, including 136 males and 64 females between 10 and 55 years old. The 3D

Fig. 6.21 *Curve* that shows standard deviation to SNR

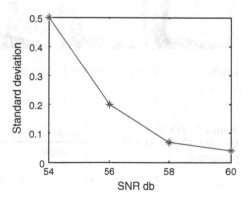

palmprint samples were collected in two separate sessions, and in each session, 10 samples were collected from both the left and right hands of the subject. The average time interval between the two sessions was one month.

Figure 6.22 shows some ROIs of 3D palmprint samples, in which each row is collected from one palm at different times. In Zhang et al. (2009), we proposed a curvature-based 3D palmprint recognition method which achieved good performance. Then, we fused 2D and 3D palmprints by an alignment refinement method (Li et al. 2010) to further improve the performance.

Table 6.2 lists the EERs by 2D, 3D and 2D + 3D palmprints, and Fig. 6.23 illustrates the corresponding ROC curves. From Table 6.2 and Fig. 6.23, we can see that with a 3D palmprint alone, the EER is 0.294 %, which is accurate enough for most of the applications, and by fusing 2D and 3D palmprints, the EER is much better than for any single pattern. These experiments demonstrate the effectiveness of the proposed 3D palmprint system.

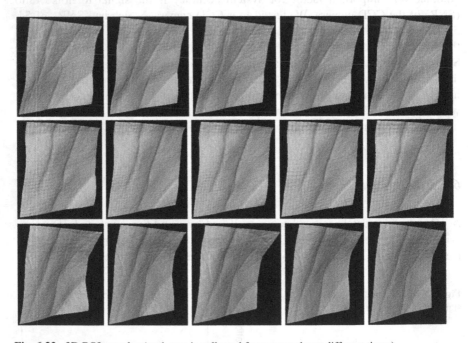

Fig. 6.22   3D ROI samples (each row is collected from one palm at different times)

Table 6.2 The EERs from 2D, 3D and 2D + 3D palmprints	Methods	EER (%)
	2D palmprint	0.046
	3D palmprint	0.294
	Fusing 2D and 3D palmprints	0.025

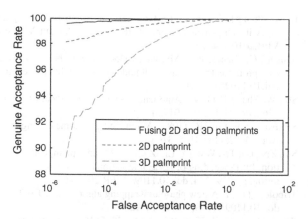

Fig. 6.23 The ROC *curves* from 2D, 3D and 2D + 3D palmprints

6.5 Summary

In this chapter, we have introduced a novel 3D palmprint acquisition system based on structured-light imaging technology. The principal and process of the generation of 3D data have been discussed. The details of the components of the 3D palmprint acquisition device have been described. The measurement accuracy of the developed system is between 0.05 and 0.1 mm which is accurate enough for capturing the depth information of palmprints. The experiments show that the developed 3D palmprint acquisition system can obtain valuable 3D information from the palmprint which is useful for palmprint identification.

References

Cyberware Inc (2010) Cyberware rapid 3D scanners. http://www.cyberware.com/. Accessed 21 May 2012

Fiona B, Paul P, James C (2003) A theoretical comparison of three fringe analysis methods for determining the three-dimensional shape of an object in the presence of noise. Opt Lasers Eng 39(1):35–50. doi:10.1016/S0143-8166(02)00071-4

Fratric I, Ribaric S (2008) Colour-based palm print verification: an experiment. In: The 14th IEEE mediterranean. doi:10.1109/MELCON.2008.4618550

Hung YY, Lin L, Shang HM, Park BG (2000) Practical three-dimensional computer vision techniques for full-field surface measurement. Opt Eng 39(1):143–149. doi:10.1117/1.602345

Huntley JM, Saldner H (1993) Temporal phase-unwrapping algorithm for automated interferogram analysis. Appl Opt 32(17):3047–3052. doi:10.1364/AO.32.003047

Jia W, Huang DS, Zhang D (2008) Palmprint verification based on robust line orientation code. Pattern Recogn 41(5):1504–1513. doi:10.1016/j.patcog.2007.10.011

Judge TR, Bryanston-Cross PJ (1994) A review of phase unwrapping techniques in fringe analysis. Opt Lasers Eng 21(4):199–239. doi:10.1016/0143-8166(94)90073-6

Kong AW, Zhang D (2004) Competitive coding scheme for palmprint verification. Int Conf Pattern Recogn 1:520–523. doi:10.1109/ICPR.2004.1334184

Konica Minolta Sensing Inc (2010) Konica Minolta VIVID 9i/910. http://www.3dscanco.com/products/3d-scanners/3d-laser-scanners/konica-minolta/. Accessed 1 July 2012

Li W, Zhang L, Zhang D, Lu GM, Yan JQ (2010) Efficient joint 2D and 3D palmprint matching with alignment refinement. IEEE Conf Comput Vision Pattern Recogn. doi:10.1109/C
VPR.2010.5540134

Sansoni G, Biancardi L, Minoni U, Docchio F (1994) A novel, adaptive system for 3-D optical profilometry using a liquid crystal light projector. IEEE Trans Instrum Meas.
doi:10.1109/19.310169

Shu W, Zhang D (1998) Automated personal identification by palmprint. Opt Eng 37(8):2359–
2362. doi:10.1117/1.601756

Srinivassan V, Liu HC (1984) Automated phase measuring profilometry of 3D diffuse object.
Appl Opt 23(18):3105–3108

Sun ZN, Tan TN, Wang Y, Li SZ (2005) Ordinal palmprint representation for personal identification. In: Proceeding of IEEE international conference on computer vision and pattern recognition, pp 279–284. doi:10.1109/CVPR.2005.267

Tribolet J (1977) A new phase unwrapping algorithm. IEEE Trans Acoust Speech Signal Process.
doi:10.1109/TASSP.1977.1162923

Wong M, Zhang D, Kong WK, Lu G (2005) Real-time palmprint acquisition system design. IEE
Proc.Vis Image Signal Process. doi:10.1049/ip-vis:20049040

Zhang D, Shu W (1999) Two novel characteristics in palmprint verification: datum point
invariance and line feature matching. Pattern Recogn 32(4):691–702. doi:10.1016/
S0031-3203(98)00117-4

Zhang D, Kong AW, You J, Wong M (2003) On-line palmprint identification. IEEE Trans Pattern
Anal Mach Intell 25(9):1041–1050. doi:10.1109/TPAMI.2003.1227981

Zhang D, Lu G, Li W, Zhang L, Luo N (2009) Palmprint Recognition Using 3-D Information.
IEEE Trans Syst Man Cybern. doi:10.1109/TSMCC.2009.2020790

Chapter 7
3D Information in Palmprint

Abstract Palmprint has been proved to be one of the most unique and stable biometrics characteristics. Almost all the current palmprint recognition techniques capture the two-dimensional image of the palm surface and use it for feature extraction and matching. Although two-dimensional palmprint recognition can achieve high accuracy, the two-dimensional palmprint images can be easily counterfeited and much three-dimensional depth information is lost in the imaging process. This chapter explores a three-dimensional palmprint recognition approach by exploiting the three-dimensional structural information of the palm surface. The structured-light imaging is used to acquire the three-dimensional palmprint data, from which several types of unique features, including Mean Curvature Image, Gauss Curvature Image and Surface Type, are extracted. A fast feature matching and score level fusion strategy are proposed for palmprint matching and classification. With the established three-dimensional palmprint database, a series of verification and identification experiments is conducted to evaluate the proposed method. The results demonstrate that three-dimensional palmprint technique has high recognition performance. Although its recognition rate is a little lower than two-dimensional palmprint recognition, three-dimenwsional palmprint recognition has higher anti-counterfeiting capability and it is more robust to illumination variations and serious scrabbling in the palm surface. Meanwhile, by fusing the two-dimensional and three-dimensional palmprint information, much higher recognition rate can be achieved.

Keywords 3D palmprint recognition • Biometrics • Feature extraction • Surface curvature

D. Zhang and G. Lu, *3D Biometrics*, DOI: 10.1007/978-1-4614-7400-5_7,
© Springer Science+Business Media New York 2013

7.1 Introduction

Biometrics authentication is playing important roles in applications of public security, access control, forensic, banking, etc. (Bolle et al. 2003). The commonly used biometrics characteristics include fingerprint, face, iris, signature, gait, etc. In the past decade, palmprint recognition has been growing rapidly, starting from the inked palmprint based off-line methods (Zhang 1999, 2004) and developing to the CCD camera based on-line methods Kong et al. 2004; Zhang 2004; Kong et al. 2006; Sun et al. 2005). Almost all the current palmprint recognition techniques and systems are based on the two dimensional (2D) palm images (inked images or CCD camera captured images). Although 2D palmprint recognition techniques can achieve high accuracy, the 2D palmprint can be easily counterfeited and much three dimensional (3D) palm structural information is lost in the 2D palmprint acquisition process. In addition, strong illumination variations and serious scrabbling in the palm may invalidate 2D palmprint recognition. Therefore, it is of high interest to explore new palmprint recognition techniques to overcome these difficulties. Intuitively, 3D palmprint recognition is a good solution.

In the on-line 2D palmprint recognition system, a CCD camera is used to acquire the palmprint image. The main features in 2D palmprint include principal lines and wrinkles. Although 2D palmprint recognition has proved to be efficient in term of verification rate, it has some inherent drawbacks. First, the palm is not a pure plane and the 3D depth information cannot be captured by using a single CCD camera. Second, the illumination variations in the system will affect a lot the 2D palmprint image and may lead to false recognition. Third, although the area of palm is large, too much contamination or too much scrabbling in the palm can still invalid the recognition. Fourth, the 2D palmprint image can be easily copied and counterfeited so that the anti-forgery ability of 2D palmprint needs improvement.

Recently, 3D techniques have been used in biometrics authentication, such as 3D face (Kakadiaris et al. 2007; Samir et al. 2006; Gokberk et al. 2008) and 3D ear recognition (Chen et al. 2007; Yan et al. 2007). Range data are usually used in these 3D biometrics applications. Most of the existing commercial 3D scanners use laser triangulation to acquire the 3D depth information, for example, the widely used Minolta VIVID Series (Samir et al. 2006; Chen et al. 2007; Yan et al. 2007). Nonetheless, the laser triangulation based 3D imaging technique has some shortcomings for the biometrics application. For instance, the resolution of 3D cloud point may not be high enough for the requirement of accuracy in biometrics authentication; on the other hand, if we want to improve the data resolution, the laser scanning speed must be decreased and the real time requirement in biometrics authentication is hard to meet. With the above considerations, we propose to use structured-light imaging (Srinivassan et al. 1984; Stockman et al. 1988; Dunn et al. 1989) to establish the 3D palmprint acquisition system. The structured-light imaging is able to accurately measure the 3D surface of an object while using less time than laser scanning. Figure 7.1 shows the 3D palmprint acquisition system developed by the Biometrics Research Centre, the Hong Kong Polytechnic University. There is a peg in the developed device serving as a control point to fix the hand. When the user put his/her palm on the system, an LED projector will generate structured-light stripes and project them to the palm. A series of grey level images of the palm with the stripes on it is captured by the CCD camera, and then the depth information of the palm surface is reconstructed from the stripe images.

Compared with other 3D biometrics characteristics, 3D palmprint has some desirable properties. For instance, compared with 3D ear recognition, the palmprint is much more convenient to collect and user friendly; compared with 3D face, the users do not need to close or block their eyes, and projecting stripes (or emitting laser light) on palm has much higher acceptability than that on face. In the data acquisition process, the palm can be easily placed so that the collected data is very stable. One disadvantage of 3D palmprint may be that the palm surface is relatively flat so that the depth information of palm is more difficult to capture than that of face or ear. However, as can be seen in this chapter, the curvature features of palm can be well captured by using the developed structured-light imaging system. With the proposed feature extraction and matching procedures, the whole 3D palmprint recognition system can reach very high performance in accuracy, speed and anti-counterfeit capability.

The Iterative Closest Point (ICP) algorithm (Besl et al. 1992) is widely used in many 3D object recognition systems for matching. The ICP schemes, however, are not suitable for 3D palmprint matching because the noise and variations of the palm may have much impact on the matching score. In this chapter, we will extract the local curvature features of 3D palmprint for classification and matching. After the 3D depth information of palm is obtained, a sub-area, called the Region of Interest (ROI), of the 3D palmprint image is extracted. Besides reducing the data size for processing, the ROI extraction process also serves to align the palmprints and normalize the area for feature extraction. The Mean curvature and Gaussian curvature features of each cloud point in the ROI are then calculated.

Fig. 7.1 The developed 3D palmprint authentication system with structured-light imaging

To save storage space and speed up the matching process, we convert the curvature features to grey level images, i.e. the Mean Curvature Image (MCI) and Gaussian Curvature Images (GCI). The third kind of 3D feature, the Surface Types (ST), of the palmprint is also defined and extracted. Finally, by fusing the MCI/GCI and ST features, the input palm can be classified and recognized. We established a 3D palmprint database with 6,000 samples from 260 people. Extensive experiments are conducted to evaluate the performance of the proposed feature extraction and matching schemes.

The rest of the chapter is organized as follows. Section 7.2 describes the acquisition of 3D palmprint data. Section 7.3 discusses the ROI region determination and the 3D feature extraction from ROI. The calculation of palm curvatures and the generation of GCI, MCI and ST features are described in detail. Section 7.4 presents the feature matching and fusion methods. Section 7.5 shows the experimental results and Sect. 7.6 concludes the chapter.

7.2 3D Palmprint Data Acquisition

The commonly used 3D imaging techniques include multi-viewpoint reconstruction (Hemandez et al. 2008; Hartley et al. 2000), laser scanning (Blais et al. 1988) and structured-light scanning (Srinivassan et al. 1984; Saldner et al. 1997; Stockman et al. 1988; Dunn et al. 1989; Hu et al. 1989; Sanderson et al. 1988). Roughly speaking, multi-viewpoint reconstruction is a low cost but low accuracy method. It is suitable to measure objects which have obvious corner features. Another problem of multi-viewpoint based 3D reconstruction is that it may be hard and time consuming to find the correspondence points in different viewpoint images. Laser scanning is a popular 3D scanning method and it is able to reach high resolution but at the cost of expensive instrument and long collecting time. Particularly, for sweating palms, laser scanning will be affected greatly by the palm surface reflection. Structured-light imaging is a well established 3D scanning technique. By using phase shifting method, it can measure the object surface in a high accuracy and in a relatively short time.

Considering the requirements of accuracy and speed in biometrics authentication, in the application of 3D palmprint recognition we choose to use structured-light scanning to acquire the palm depth information. The use of structured-light in object surface measurement can be traced back to more than 2 decades ago (Srinivassan et al. 1984). Since then, it has been widely used in many applications, such as 3D object measurement, 3D shape reconstruction, reverse engineering, etc. (Stockman et al. 1988; Dunn et al. 1989; Hu et al. 1989). In structured-light imaging, a light source projects some structured light patterns (stripes) onto the surface of the object. The reflected light is captured by a CCD camera and then a series of images is collected. After some calculation, the 3D surface depth information of the object can be obtained. In the earlier stage, parallel light such as laser

(Sanderson et al. 1988) or point light array (Stockman et al. 1988) were used. With the development of the light source techniques, liquid crystal light projectors have been successfully used as the light source (Sansoni et al. 1994). In our developed system, a cost-effective grey LCD projector with LED light source is employed, and some shift light patterns are projected to the palm.

Figure 6.7 shows the architecture of the developed 3D palmprint data acquisition device. It is mainly composed of a light projecting unit and a data collection unit. The light projecting unit mainly contains a white LED light source, an LCD panel and several convergent lenses. The data collection unit contains a CCD camera and a camera lens. Figure 6.11 illustrates the 3D palmprint data collection and processing process. The computer controls a projector to project a series of 13 structured-light stripes to the palm inner surface and the CCD camera captures the palm images with projected stripes on it. At the same time the computer sends a command to the data collection board to store the images. The data collection costs about 2 s. From these palm images, the depth information of each point on the palm can be computed using phase transition and phase expansion techniques (Saldner et al. 1997). These processing, which are marked using blue arrows in Fig. 6.11, will cost about 0.5 s. So the total time for 3D palmprint generation is about 2.5 s.

Figure 6.5 illustrates the imaging principle of the structured-light technique (Srinivassan et al. 1984). Interested readers can refer to (Srinivassan et al. 1984) for more details about structured-light imaging. In Fig. 7.2, there is a reference plane whose height is 0. By projecting light through grating to the object surface,

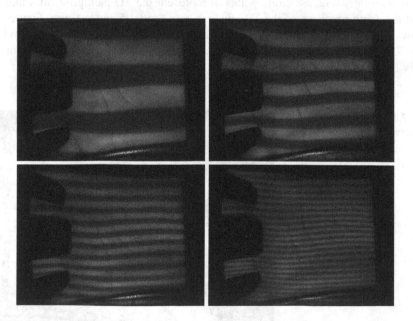

Fig. 7.2 Sample patterns of the stripes on the palm

the relative height of a point D at spatial position (x, y) to the reference plane can be calculated as follows (Srinivassan et al. 1984)

$$h(x,y) = \overline{BD} = \frac{\overline{AC} \cdot \tan \theta_0}{1 + \tan \theta_0 / \tan \theta_n}, \tag{7.1}$$

with

$$\overline{AC} = \frac{\phi_{CD}}{2\pi} P_0, \tag{7.2}$$

where P_0 is the wavelength of the projected light on the reference plane, θ_0 is the projecting angle, θ_n is the angle between the reference plane and the line which passes through the current point and the CCD center, and ϕ_{CD} is the phase difference between points C and D. Because the phase of point D on the 3D object is equal to the phase of point A on the reference plane, ϕ_{CD} can be calculated as:

$$\phi_{CD} = \phi_{CA} = \phi_{OC} - \phi_{OA}. \tag{7.3}$$

By using Eq. 7.1 and the phase shifting and unwrapping technique (Saldner et al. 1997), we can retrieve the depth information of the object surface by projecting a series of phase stripes on it (13 stripes are used in our system). Some sample patterns of the stripes on the palm are illustrated in Fig. 7.3.

With the above processing, the relative height of each point, $h(x,y)$, could be calculated. The range data of the palm surface can then be obtained. In the developed system, the size of the 3D image is 768×576 with 150 dpi resolution, i.e. there are totally 442,368 cloud points to represent the 3D palmprint information. And the depth precision of the 3D image is between 0.05 and 0.1 mm. Figure 7.3 shows an example 3D palmprint image captured by the system. The gray level in Fig. 7.3 is related to the value of $h(x,y)$ and it is rendered by OpenGL automatically for better visualization.

Fig. 7.3 An example of captured 3D palmprint image

7.3   Feature Extraction from 3D Palmprint

7.3.1   ROI Extraction

From Fig. 7.4, we can see that in the 3D palmprint image of resolution 768 × 576, many cloud points, such as those in the boundary area and those in the fingers, could not be used in feature extraction and recognition. Most of the useful and stable features locate in the center area of the palm. In addition, at different times when the user puts his/her hand on the system, there will be some relative displacements of the positions of the palm, even that we impose some constraints on the users to place their hands. Therefore, before feature extraction it is necessary to perform some preprocessing to align the palmprint and extract the central area of it, which is called the Region of Interest (ROI).

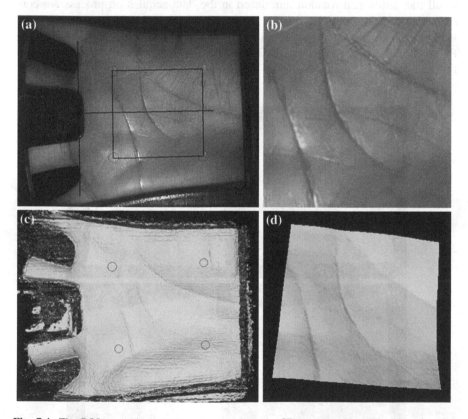

Fig. 7.4 The ROI extraction of 3D palmprint from its 2D counterpart. **a** The 2D palmprint image, the adaptively established coordinate system and the ROI (i.e. *the rectangle*). **b** The extracted 2D ROI. **c** The 3D palmprint image, whose cloud points have a one-to-one correspondence to the pixels in the 2D counterpart. **d** The obtained 3D ROI by grouping the cloud points corresponding to the pixels in 2D ROI. (Please note that we change the viewpoint for better visualization.)

Using the developed structured-light based 3D imaging system, the 2D and 3D palmprint images can be obtained simultaneously, and there is a one-to-one correspondence between the 3D cloud points and the 2D pixels. Therefore, the ROI extraction of the 3D palmprint data can be easily implemented via the 2D palmprint ROI extraction procedure. In this chapter, we use the algorithm in to extract the 2D ROI. Once the 2D ROI is extracted, the 3D ROI is obtained by grouping the cloud points that are in correspondence to the pixels in the 2D ROI. Figure 7.4 illustrates the ROI extraction process. Figure 7.4a shows a 2D palmprint image, the established local coordinate system by using the algorithm in and the ROI (i.e. the rectangle); Fig. 7.4b shows the extracted 2D ROI; Fig. 7.4c shows the 3D palmprint image and Fig. 7.4d shows the obtained 3D ROI by grouping the cloud points corresponding to the pixels in 2D ROI. (Please note that we change the viewpoint for better visualization)

By using ROI extraction procedure, the 3D palmprints are aligned so that the small translation and rotation introduced in the data acquisition process are corrected. In addition, the data amount used in the following feature extraction and matching process is significantly reduced. This will save much computational cost. Figure 7.5a shows the extracted 3D ROIs of two palmprints from the same person. Figure 7.5b shows another example.

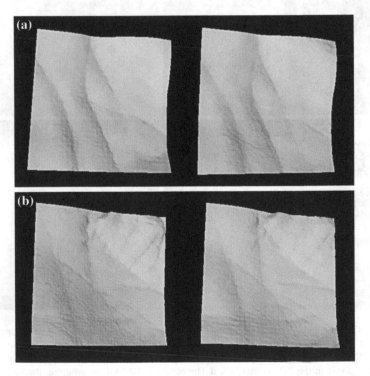

Fig. 7.5 a The 3D ROIs extracted from two palmprints of one person. **b** The 3D ROIs extracted from two palmprints of another person

7.3.2 Curvature Calculation

With the obtained ROI, stable and unique features are expected to be extracted for the following pattern matching and recognition. The depth information in the acquired 3D palmprint reflects the relative distance between the reference plane and each point in the object (referring to Fig. 7.4). As can be seen in Fig. 7.1, most of the palm region, which will be captured by the system for personal recognition, will not touch the device. Therefore, there is little pressure on the palm. In data acquisition, the users are asked to put their hands naturally on the device. Thus, the pressure-caused deformation of the palmprint image is actually very small. The z-values of the 3D cloud points are mainly affected by noise and the pose change of hand in scanning. The ROI extraction process can only correct the rotation and translation displacements in the x–y plane but not the z-axis. The noise and variations in the 3D palmprint cloud points make the well-known ICP algorithms (Besl et al. 1992) not suitable for 3D palmprint recognition. Instead, the local invariant features, such as the curvatures (Kühnel 2006) of the principal lines and strong wrinkles in the palm surface, will be much more stable in representing the characteristics of 3D palmprint.

Let p be a point on the surface S. Consider all curves C_i on S passing through the point p. Each curve C_i will have an associated curvature K_i at p. Among those curvatures K_i, at least one is characterized as maximal k_1 and one as minimal k_2, and these two curvatures k_1 and k_2 are known as the principal curvatures of point p on the surface (Kühnel 2006). The Mean curvature H and the Gaussian curvature K of p are defined as follows

$$H = \frac{1}{2}(k_1 + k_2), \ K = k_1 * k_2. \tag{7.4}$$

The Mean and Gaussian curvatures are intrinsic measures of a surface, i.e. they depend only on the surface shape but not on the way how the surface is placed in the 3D space (Kühnel 2006). Thus, such curvature features are robust to the rotation, translation and even some deformation of the palm. The captured 3D palmprint data are organized range data. We adopt the algorithm in (Besl et al. 1988) to estimate the Mean and Gaussian curvatures for its simplicity and effectiveness. For more information, please refer to (Besl et al. 1988).

7.3.3 Mean Curvature Image and Gaussian Curvature Image

With the method in (Besl et al. 1988), the Mean and Gaussian curvatures of a 3D palmprint image can be calculated. Figure 7.6 shows the distribution of the Mean and Gaussian curvatures of 100 3D palmprint images.

For better visualization and more efficient computation, we convert the original curvature images into grey level integer images according to the distributions. We

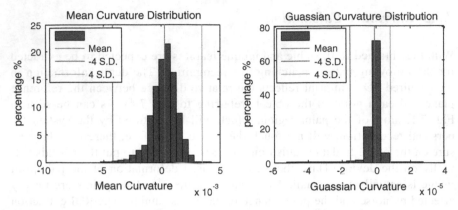

Fig. 7.6 Distributions of Mean and Gaussian curvatures of 3D palmprint images

first transform the curvature image C (Gaussian curvature K or Mean curvature H) into $\bar{C}$ as follows

$$\bar{C}(i,j) = 0.5(C(i,j) - \mu)/(4\delta) + 0.5, \tag{7.5}$$

where μ is the mean of the curvature image. With Eq. 7.5, most of the curvature values will be normalized into the interval [0,1]. We then map $\bar{C}(i,j)$ to an 8-bits grey level image $G(i,j)$:

$$G(i,j) = \begin{cases} 0 & \bar{C}(i,j) \leq 0 \\ round\,(255 \times \bar{C}(i,j)) & 0 < \bar{C}(i,j) < 1. \\ 255 & \bar{C}(i,j) \geq 1 \end{cases} \tag{7.6}$$

We call images $G(i,j)$ the Mean Curvature Image (MCI) and Gaussian Curvature Image (GCI), respectively for Mean and Gaussian curvatures. Figure 7.7 illustrates the MCI and GCI images of three different palms and Fig. 7.8 illustrates the MCI and GCI images of a palm at different acquisition times. We can see that the 2D MCI and GCI images can well preserve the 3D palm surface features. Not only the principal lines, which are the most important texture features in palmprint recognition, are clearly enhanced in MCI/GCI, but also the depth information of different shape structures is well preserved. The MCI/GCI feature images provide us a good basis for further processing and pattern matching.

7.3.4 Points Classification Using Surface Types

Since the palm is a continuous surface with different convex and concave structures, we can classify the points in the palm into different groups based on their local surface characteristics. Such kind of 3D feature is called Surface Type

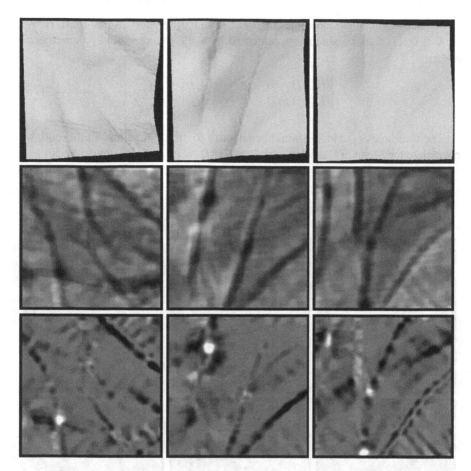

Fig. 7.7 The 3D ROI images (*first row*) of three different palmprints and their MCI (*second row*) and GCI (*third row*) images. From *left* to *right*, each column shows the images of one palm, respectively

(ST), and it can be determined by the signs of Mean and Gaussian curvature values (Woodard et al. 2005). In (Besl et al. 1988), eight fundamental surface types (ST) were defined. Figure 7.9 illustrates the shapes of the eight STs and Table 7.1 lists the definition of them based on the corresponding Mean Curvature H and Gaussian Curvature K. In total nine STs can be defined, including the eight fundamental STs in Fig. 7.9 and another special case for $H = 0$ and $K > 0$. Using STs, the points in the 3D palmprints can be intuitively classified into nine classes. For example, the GCI images in Figs. 7.10 and 7.11 have some white pixels, which implies $K > 0$; meanwhile, the corresponding points in the MCI images are black, which implies $H < 0$. From Table 7.1, we know that these points belong to the PEAK surface type ($ST = 1$).

We need to quantize the values of H and K to fix the intervals that make $H = 0$ or $K = 0$. This can be simply implemented by using two threshold parameters ε_H and ε_K:

$$\begin{cases} H(i,j) = 0 & if \quad |H(i,j)| < \varepsilon_H \\ K(i,j) = 0 & if \quad |K(i,j)| < \varepsilon_K. \end{cases} \tag{7.7}$$

The thresholds ε_H and ε_K should be adaptive to different palms. To this end, we normalize the Mean or Gaussian curvature $C(i,j)$ by its standard deviation as

$$C_s(i,j) = C(i,j)/2\delta. \tag{7.8}$$

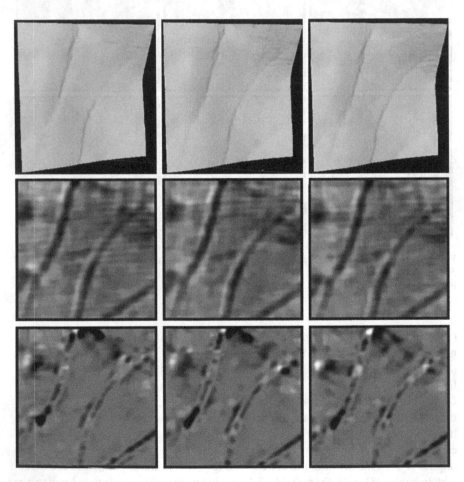

Fig. 7.8 The 3D ROI images (*first row*) of the same palmprint but collected at different times and their MCI (*second row*) and GCI (*third row*) images. From *left* to *right*, each column shows the images for each time, respectively

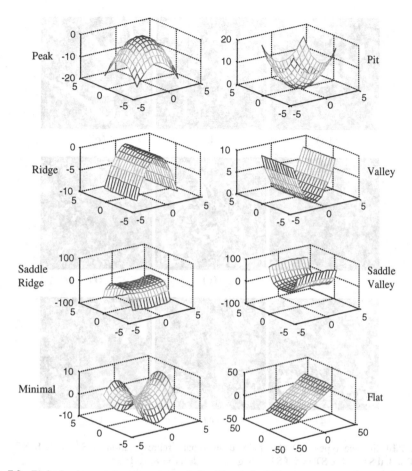

Fig. 7.9 *Eight* fundamental surface types defined by their different convex and concave structures (Woodard et al. 2005)

Table 7.1 Surface type labels from curvature signs

	K > 0	K = 0	K < 0
H < 0	Peak	Ridge	Saddle ridge
	ST = 1	ST = 2	ST = 3
H = 0	None	Flat	Minimal surface
	ST = 4	ST = 5	ST = 6
H > 0	Pit	Valley	Saddle valley
	ST = 7	ST = 8	ST = 9

Using Eq. 7.8, most curvature values will fall into the interval $[-1, 1]$ without changing their signs. Then, we can easily set the thresholds ε_H and ε_K around zero in Eq. 7.7.

With the above procedures, each point in the 3D palmprint can be classified into one of the nine surface types. Figure 7.10 shows the classification results of a

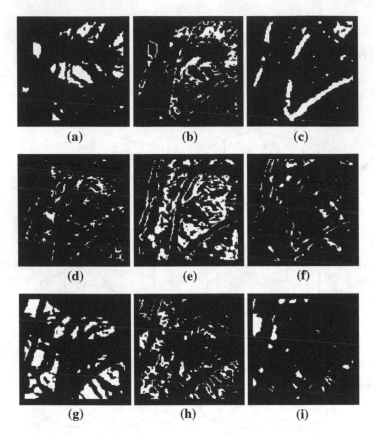

Fig. 7.10 Surface types 1–9 of a palm using binary representation. **a** ST = 1. **b** ST = 2. **c** ST = 3. **d** ST = 4. **e** ST = 5. **f** ST = 6. **g** ST = 7. **h** ST = 8. **i** ST = 9

palm sample using binary images, i.e. a white pixel in an ST image means that it belongs to the corresponding ST. These features can be used separately or jointly in the matching process.

7.4 Feature Matching and Fusion

Using the techniques developed in Sect. 7.3, the MCI, GCI and ST features of the 3D palmprint can be extracted. The different palmprints can then be matched based on those features and the recognition can be accomplished according to the matching score. In this section, we discuss the matching strategy of different features and the fusion of them. We will then discuss the fusion of 2D and 3D palmprint information for higher recognition rates.

7.4.1 MCI/GCI Feature Matching

The extracted MCI/GCI features (referring to Eq. 7.6) are 8-bits grey level images. The variations of the local curvatures are mainly caused by acquisition noise and the pose changes of the hand. To reduce such variations and extract the stable and intrinsic curvature features, we binarize the MCI and GCI maps in the following processing. The principal lines and strong wrinkles are the most stable and significant features in the palmprint. After the binarization, they could be well preserved in the MCI and GCI maps, while the noise caused small variations are removed. In addition, the binarization of the MCI and GCI maps can make the feature matching very fast. Here we simply convert the MCI/GCI into binary images by using adaptive threshold:

$$B(i,j) = \begin{cases} 1 & G(i,j) < c \cdot \mu_G \\ 0 & others \end{cases}, \tag{7.9}$$

where c is a constant and μ_G is the mean value of $G(i,j)$. With our experimental experience, we set $c = 0.7$ in the experiments. Figure 7.11 shows the binarized versions of the MCI/GCI images in Fig. 7.10.

We use the AND operation to calculate the matching score of MCI/GCI features. Denote by B_d the binarized MCI/GCI image in the database and by B_t the

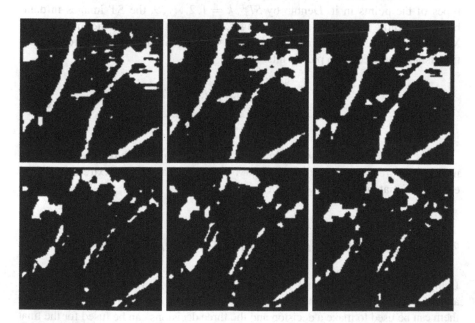

Fig. 7.11 The binarized MCI images (*first row*) and GCI images (*second row*)

input MCI/GCI binary image. Suppose the image size is $n \times m$. The matching score between B_d and B_t is defined as:

$$R_C = \frac{2 \sum\limits_{i=1}^{n} \sum\limits_{j=1}^{m} B_d(i,j) \oplus B_t(i,j)}{\sum\limits_{i=1}^{n} \sum\limits_{j=1}^{m} B_d(i,j) + \sum\limits_{i=1}^{n} \sum\limits_{j=1}^{m} B_t(i,j)}, \tag{7.10}$$

where symbol "$\oplus$" means logic operation AND. If the two MCI/GCI binary feature images B_d and B_t are the same, then we have $R_c = 1$; the minimum value of R_c is 0, which means that the two binary images have no overlap "1" pixel.

Since there may still have some displacements between the two palmprint images even after ROI extraction, when calculating the matching score by Eq. 7.10, we will shift two, four, six and eight pixels of the test image along 8 directions: right, left, up, down, left-up, left-down, right-up and right-down, respectively. Thus we will have $8 \times 4 + 1 = 33$ matching scores and the maximum one is selected.

7.4.2 ST Feature Matching

For each 3D palmprint, we have 9 binary ST images, representing different surface types of the points in it. Denote by ST_k^d, $k = 1, 2, \ldots, 9$, the ST images in database and denote by ST_k^t the test ST images. Different from the matching score of MCI/GCI features, here we use the absolute value of difference (AVD) to measure the distance between two palmprints:

$$R_{ST} = 1 - \frac{\sum\limits_{k=1}^{9} \sum\limits_{i=1}^{n} \sum\limits_{j=1}^{m} |ST_k^d(i,j) - ST_k^t(i,j)|}{2 \times m \times n}. \tag{7.11}$$

If the ST features of two palmprints are identical, we have the maximum matching score $R_{ST} = 1$. On the contrary, if the ST features are extremely different, the ST matching score will be $R_{ST} = 0$.

7.4.3 Matching Score Fusion of MCI, GCI and ST

Using equations Eqs. 7.10 and 7.11, three matching scores (two R_c scores for MCI and GCI respectively, and one R_{ST} score for ST) can be fast calculated. Each one of them can be used to make a decision and the three decisions can be fused for the final decision. Another way is to fuse the three matching scores first and then the decision is made based on the fused matching score. Here we adopt the second strategy.

Suppose there are n matching scores and denote them by R_i, $i = 1, 2, \ldots, n$. The commonly used score level fusion techniques include Min-Score(MIN) $R_{MIN} = \min(R_1, R_2, \ldots, R_n)$, Max-Score(MAX) $R_{MAX} = \max(R_1, R_2, \ldots, R_n)$, Summation (SUM) $R_{SUM} = \frac{1}{n} \sum_{i=1}^{n} R_i$ and Weighted Average (WA) methods (Snelick et al. 2005; Indovina et al. 2003). We can see that the Mean curvature contains more useful information than Gaussian curvature so that the MCI feature can lead to better matching result than GCI (referring to Sect. 7.5.3). Therefore, we should assign a greater weight to MCI than to GCI. Because the Equal Error Rate (EER) is an important index of the matching result and it can be estimated by test database, the weights can be determined according to the corresponding EER values.

In (Snelick et al. 2005), a WA scheme, called Matcher Weighting (MW), is proposed:

$$R_{MW} = \sum_{i=1}^{n} w_i R_i, \quad w_i = \frac{1/e_i}{\sum_{j=1}^{n} 1/e_i}, \quad (i = 1, 2 \ldots, n), \tag{7.12}$$

where w_i is the weight of R_i, and e_i is the corresponding EER. The MW scheme assigns smaller weights to those features with higher EER values. In (Snelick et al. 2005), a user weighting (UW) fusion scheme was also proposed. The idea comes from the wolf-lamb concept introduced by (Doddington et al. 1998). The UW method assigns different weights to different users of different matchers according to the user's inimitable property. The user who can be imitated easily is called as lamb and will be assigned with lower weight. However, in practice whether the user is a lamb or not highly depends on the database, which limits the application of UW. In our experiments in Sect. 7.5, we will test the MIN, MAX, SUM and MW fusing methods.

7.4.4 Fusion of 2D and 3D Palmprint Information

Intuitively, the 2D and 3D palmprint information can be fused for better recognition accuracy. The fusion can be performed on either matching score level or feature level. For score level fusion, we regard each pair of 2D and 3D matching scores as a two dimension vector, and then use the linear SVM method (Vapnik 1995) to classify the genuine and impostor. Besides score level fusion, feature level fusion is also applicable to 2D and 3D palmprint to improve the matching performance. The main features in 2D palmprint are principal lines and wrinkles, and these features can also be represented by 3D palmprint. Sometimes the key features in 2D image may not be well captured (e.g. over-illumination) but these critical discriminant features can be enhanced by fusing with the 3D information. Here we use an example shown in Fig. 7.12 to illustrate the proposed feature level fusion scheme. Fig. 7.12a is a 2D ROI which is over-illuminated; Fig. 7.12b is the

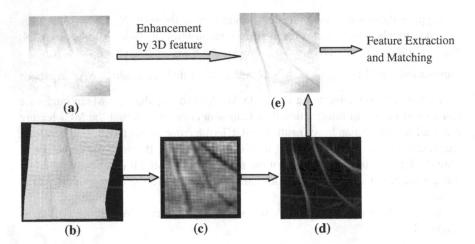

Fig. 7.12 Illustration of the 2D and 3D feature level fusion. **a** 2D ROI. **b** 3D ROI. **c** MCI. **d** Energy image of MCI. **e** Enhanced 2D ROI

corresponding 3D ROI; Fig. 7.12c is the MCI feature extracted from the 3D ROI; Fig. 7.12d is the energy image of the MCI; and Fig. 7.12e is the enhanced 2D ROI by the 3D MCI energy image. Here we use the method in (Huang et al. 2008) to calculate the energy image of the MCI map.

Denote by H the $m \times n$ MCI map. Then the energy image of H, denoted by E, is calculated as

$$E(i,j) = \min \left\{ e_{i,j}^1, \dots, e_{i,j}^k, \dots, e_{i,j}^K \right\}, \quad i = 1, \dots, m; \quad j = 1, \dots, n, \quad (7.13)$$

with

$$e_{i,j}^k = \sum_{(u,v) \in L_{i,j}^k} H(u,v), \quad (7.14)$$

where $L_{i,j}^k$ is a predefined template centered at (i, j) and K is the total number of templates. Some templates used in this chapter are given in Fig. 7.13. They are basically the templates to model the line structures of different orientations. After calculating the energy image E, we enhance the 2D ROI, denoted by F, as follows:

$$F_E(i,j) = F(i,j) - \alpha \cdot E(i,j), \quad (7.15)$$

where α is a parameter to control the fusion and we set $\alpha = 0.3$ in our experiments. In Fig. 7.12e, we can see that the line features are much improved after fusing with the MCI map. They will be well extracted in the following 2D palmprint feature extraction process and will lead to more robust matching results. Our experimental results in Sect. 7.5.3 validate this.

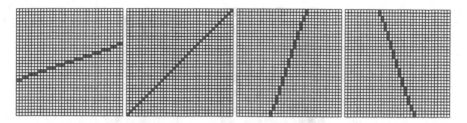

Fig. 7.13 Examples of *line* structure templates

7.5 Experimental Results

7.5.1 Anti-Counterfeiting Test

The proposed 3D palmprint recognition system uses structured-light technique to capture the palm surface depth information. One clear advantage of 3D palmprint recognition over 2D palmprint is that it has higher anti-counterfeiting capability because a 3D palm is much harder to forge. In this section we made anti-counterfeiting tests by using printed 2D palms and 3D palms made of plaster and olefin. A user is registered in both the 2D palmprint and 3D palmprint systems.

We first printed out a 2D palmprint image of this user on a piece of paper using a high quality laser printer (HP LaserJet 1,020 plus, 1,200 dpi). Then we put this paper on the two systems as input palm to see if it can pass the systems. Figures 7.17 and 7.18 illustrate the test and the recognition results. For 2D system, we used the competitive coding method (Kong et al. 2004). In Fig. 7.14 we can see that because the printed palm is a pure paper plane and the real palm is a curved surface, the captured palmprint images from printed palm and real palm will have some differences on illumination level. However, the features (principal lines, wrinkles, etc.) of the two images are identical and the system will successfully match the printed palm with the palmprints in the database because the 2D palmprint recognition algorithms can deal with the illumination variations to some extent. As what we expected, in Fig. 7.15 it can be seen that 3D palmprint technique can easily tell the counterfeited palm because the there is no depth variation information in the printed paper and the captured 3D palmprint image from printed 2D palmprint is just a noise image.

We made two kinds of 3D counterfeiting palms by plaster and olefin, respectively, to test whether they can break the 3D palmprint system. For the plaster counterfeiting palm, we first press a real palm on plasticine to record the palm depth information, and then mould the liquid plaster into the plasticine model to make the 3D palmprint counterfeit, which is shown in Fig. 7.16. The anti-counterfeit test on this plaster palm is illustrated in Figs. 7.20 and 7.21. We can see that both the 2D and 3D systems reject the plaster palm. The 2D system rejects the fake palm because the line features captured in the 2D image are weak,

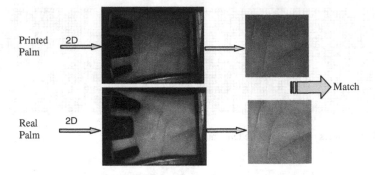

Fig. 7.14 Anti-counterfeiting test of printed palm using 2D palmprint recognition

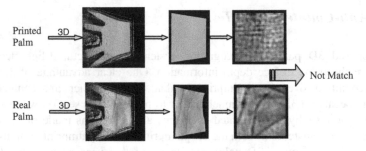

Fig. 7.15 Anti-counterfeiting test of printed palm using 3D palmprint recognition

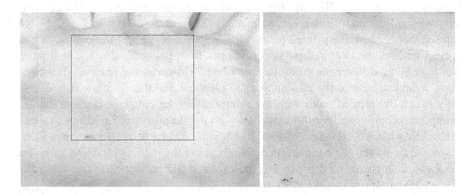

Fig. 7.16 A 3D plaster palmprint

as shown in Fig. 7.17. The 3D system also rejects the fake palm. This is because there is some deformation on the palm surface in the making of the plaster palm. Therefore, the extracted curvature features of the plaster palmprint are very different from those of the real palmprint, as illustrated in Fig. 7.18. Similar results have been gotten for the counterfeiting test on the olefin palm because it's hard to

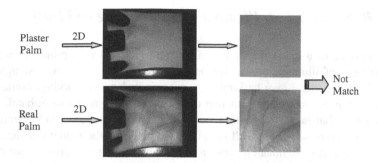

Fig. 7.17 Anti-counterfeiting test of plaster palm using 2D palmprint recognition

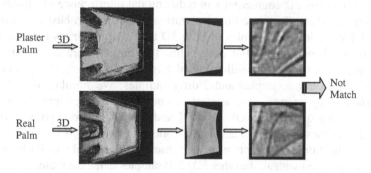

Fig. 7.18 Anti-counterfeiting test of plaster palm using 3D palmprint recognition

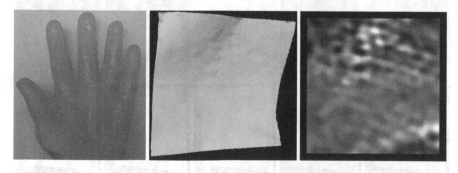

Fig. 7.19 The counterfeiting palm made by olefin. From *left* to *right*: the counterfeiting palm, 3D ROI and the MCI of the counterfeiting palm

counterfeit the detail 3D information of the real palm. Figure 7.19 shows the olefin counterfeiting palm and its 3D ROI and MCI map. From these tests, we can see that 3D system is hard to be spoofed. Certainly, people can make a more "real" 3D palm to break the system but the cost will be much higher.

7.5.2 Robustness Test: Illumination, Scrabbling and Dirty

It is necessary to test the robustness of the proposed 3D palmprint technique to the variations of illumination, scrabbling and dirty in the palm. For example, the palm of labor workers may be seriously contaminated when working. Sometimes, one may write down some information on the palm when no paper at hand. It has been reported that in 2D palmprint recognition, light to medium level illumination variation and palm scrabbling will not affect too much on the recognition accuracy. However, when the illumination varies too much or the palm is seriously scrabbled, we can imagine that 2D palmprint will not work anymore because the 2D palmprint features (principal lines, wrinkles, etc.) are hard to capture.

Due to the different imaging principles, 3D palmprint technique has clear advantages over its 2D counterparts in reducing the interference of illumination, scrabbling and dirty. To prove this declaration, we build a robustness test database which contains 400 samples of both 3D and 2D palmprints collected from 40 palms of 20 individuals. For each palm, 10 samples were collected: 2 normal samples, 2 severely under-illuminated samples, 2 severely over-illuminated samples, 2 scrabbling samples and 2 dirty samples. We simulate the illumination variations by changing the aperture value of the camera lens. Figure 7.20 shows examples of the 2D ROI, 3D ROI and 3D MCI of one palm under the five different cases and Table 7.2 lists the collecting conditions of the five cases.

Then verification experiments were performed on this database. Each 3D (2D) sample was matched with all the other 3D (2D) samples in the database. A successful matching is called intra-class matching or genuine if the two samples are from the

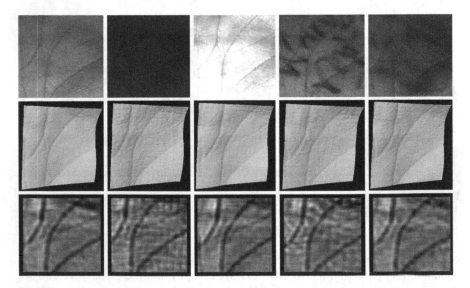

Fig. 7.20 Sample images in the robustness test database. From *top* to *bottom*: 2D ROI, 3D ROI and the 3D MCI map. From *left* to *right* are the five test cases: normal, severely under illumination, severely over illumination, scrabbling and dirty

Table 7.2 The collecting conditions of the five test cases

	Aperture value	Scrabbling	Clean/dirty
Normal	f 5.6	No	Clean
Severely under illumination	f 11	No	Clean
Severely over illumination	f 2.8	No	Clean
Scrabbling	f 5.6	Yes	Clean
Dirty	f 5.6	No	Dirty

Table 7.3 The EER of 2D and 3D palmprint verification on the robustness test database

	2D	3D
EER	5.478 %	0.344 %

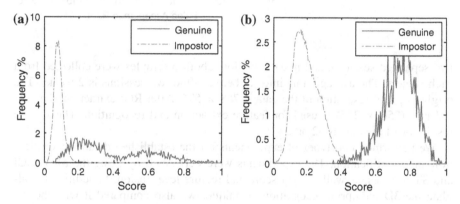

Fig. 7.21 Genuine and imposter distributions on the robustness test database. **a** 2D score distribution. **b** 3D score distribution

same class. Otherwise, the unsuccessful matching is called inter-class matching or impostor. Using the established database, there are 79,800 matching in total. The 3D matching score is obtained by fusing the MCI, GCI and ST at score level (referring to Sect. 7.4.3). The 2D matching score is calculated by using the competitive coding method (Kong et al. 2004). Table 7.3 lists the EER values of 2D and 3D palmprint verification. Figure 7.21 shows the distributions of 2D and 3D matching scores, and Fig. 7.22 shows the ROC curves. We can see that 3D information is much more robust to illumination variation, scrabbling and dirty than 2D information.

7.5.3 Database Establishment and Recognition Results

A 3D palmprint database has been established by using the developed 3D palmprint imaging device. The database contains 6,000 samples from 260 volunteers, including 182 males and 78 females. The 3D palmprint samples were collected in

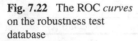

Fig. 7.22 The ROC *curves* on the robustness test database

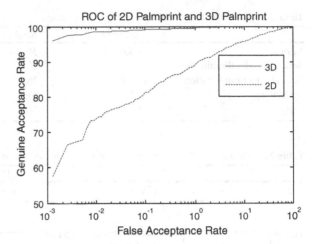

two separated sessions, and in each session, about 6 samples were collected from each subject. The average time interval between the two sessions is 2 weeks. The original spatial resolution of the data is 768 × 576. After ROI extraction, the central part (256 × 256) is used for feature extraction and recognition. The z-value resolution of the data is 32 bits.

We performed two types of experiments on the established database: verification and identification. The experiments were performed by using the MCI, GCI and ST features, as well as their score and feature level fusion. In addition to validate the 3D palmprint recognition technique, we also compared it with the 2D palmprint recognition method by using the 2D palmprint images simultaneously collected in the 3D palmprint acquisition process. The 2D palmprint verification was performed by using the competitive coding method (Kong et al. 2004). As we can see later in the experimental results, by fusing the 2D and 3D palmprint features, the highest recognition rate can be obtained.

Figure 7.23 illustrates the genuine and imposter distributions of the verification results by 3D palmprint recognition. Figure 7.26a–c show the curves by MCI, GCI and ST features respectively. Fig. 7.23d shows the curves by using the WM score level fusion of them. The receiver operating characteristic (ROC) curves by using the MCI, GCI and ST features and their fusions (MAX, MIN, SUM and MW) are shown in Fig. 7.24. The EER values, which are important indices of the verification performance, are listed in Table 7.4, where the feature extraction and matching time by using different features are also listed.[1] We see that the WM fusion achieves the smallest EER among all the schemes. From Fig. 7.24 and Table 7.4,

---

[1] The experiments were performed using Visual C++ 6.0 on a PC with Windows XP Professional, Pentium 4 CPU of 2.66 GHz and 1 GB RAM.

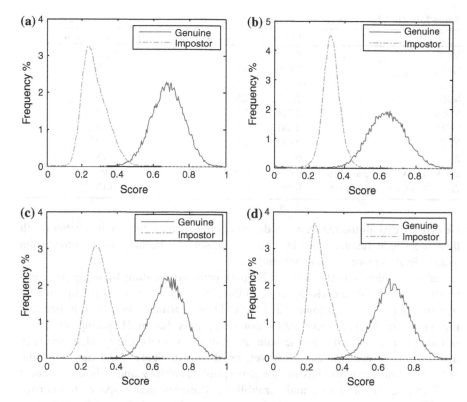

Fig. 7.23 Genuine and imposter distributions by MCI, GCI, ST features and the WM fusion of them. **a** MCI. **b** GCI. **c** ST. **d** WM fusion

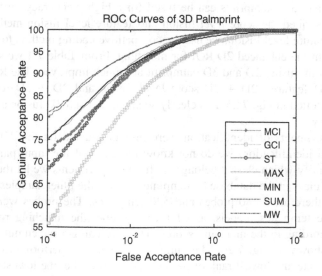

Fig. 7.24 The ROC *curves* by different matching methods

Table 7.4 The EER values, feature extraction time and matching time by different methods

	EER (%)	Feature extraction time (ms)	Matching time (ms)
MCI	0.825	112	0.86
GCI	2.059	112	0.86
ST	0.791	115	3.76
MIN	1.277	125	4.89
MAX	0.714	125	4.89
SUM	0.681	125	4.89
MW	0.632	125	4.89
2D	0.126	97	0.15
2D + 3D SVM score level fusion	0.068	189	5.01
2D + 3D feature level fusion	0.059	245	0.15

we can see that all the fusion methods, except MIN, can achieve much better result than the single feature matching. This is reasonable because more information usually lead to more accurate recognition.

Table 7.4 also lists the EER of 2D palmprint verification by using the competitive coding method (Kong et al. 2004). We can see that 2D palmprint recognition achieves much lower EER than 3D palmprint. This is mainly because the quality of 3D palmprint data is not as good as that of 2D palmprint. There is much corrupted noise in the data acquisition process and z-value accuracy needs further improvement. However, as shown in Sects. 7.5.1 and 7.5.2 the 3D palmprint technique has higher anti-spoof capability and it is more robust to illumination variations and scrabbling. There is much space to improve the 3D palmprint data acquisition precision and the performance of feature extraction and matching algorithms. The 3D palmprint recognition has great potentials.

The 2D and 3D palmprints can be fused for a higher accuracy. In Sect. 7.4.3, we have described the score level fusion and feature level fusion methods. After 2D + 3D feature level fusion, we use the competitive coding method to extract the features from the enhanced 2D ROI for matching. From Table 7.4 we see that the EER valued by fusing 2D and 3D palmprint are much improved. The ROC curves by using 2D features, 2D + 3D score level fusion and 2D + 3D feature level fusion are plotted in Fig. 7.25. It is clearly seen that 2D and 3D fusion achieves the best accuracy.

The experiments of identification were also conducted on the 3D palmprint database. In identification, we do not know the class of the input palmprint but want to identify which class it belongs to. In the experiments we let the first sample of each class in the database be template and use the other samples as probes. Therefore, there are 5,480 probes and 520 templates. The probes were matched with all the templates models, and for each probe, the matching results were ordered according to the matching scores. Then we can get the cumulative match curves as shown in Fig. 7.26. The cumulative matching performance, rank-one recognition rate and lowest rank of perfect recognition (i.e. the lowest rank when

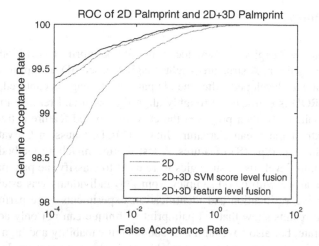

Fig. 7.25   The ROC *curves* by 2D palmprint and 2D + 3D palmprint

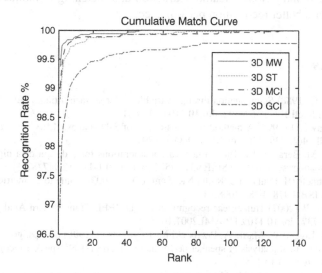

Fig. 7.26   The CMC *curves* by different matching methods in identification

Table 7.5   Cumulative matching performance by different types of features

	MCI	GCI	ST	MW
Rank-one recognition rate (%)	99.03	96.95	98.81	99.38
Lowest rank for perfect recognition	129	473	106	47

the recognition rate reaches 100 %), are listed in Table 7.5. From Fig. 7.26 we see that the performance of MW fusion is much better than other schemes.

7.6 Summary

In this chapter, we explored a new technique for palmprint based biometrics: 3D palmprint recognition. A structured-light imaging based 3D palmprint data acquisition system was developed. After the 3D palmprint image is captured, the region of interest (ROI) is extracted to roughly align the palm and remove the unnecessary cloud points. We then proposed the curvature based feature extraction algorithms to extract the Mean Curvature Image (MCI), Gaussian Curvature Image (GCI) and Surface Type (ST) features. A fast feature matching method and score level and feature level fusion strategies were used to classify the palmprints. A 3D palmprint database with 6,000 samples from 260 individuals was established, on which extensive verification and identification experiments were performed. The experimental results show that 3D palmprint technique can not only achieve high recognition rate, but also have high anti-counterfeit capability and high robustness to illumination variations and serious scrabbling in the palm surface. In the future, more advanced and powerful feature extraction and matching techniques are to be developed for a better recognition performance.

References

Besl PJ, Jain RC (1988) Segmentation through variable-order surface fitting. IEEE Trans Pattern Anal Mach Intell 10(2):167–192. doi:10.1109/34.3881

Besl PJ, McKay ND (1992) A method for registration of 3-D shapes. IEEE Trans Pattern Anal Mach Intell 14(2):239–256. doi:10.1109/34.121791

Blais F, Rious M, Beraldin JA (1988) Practical considerations for a design of a high precision 3-D laser scanner system. Proc SPIE 959:225–246. doi:10.1-117/12.947787

Bolle RM, Connell JH, Pankanti S, Ratha NK, Senior AW (2003) Guide to biometrics. Springer, New York. ISBN 978-0-387-40089-1

Chen H, Bhanu B (2007) Human ear recognition in 3D. IEEE Trans Pattern Anal Mach Intell 29(4):718–737. doi:10.1109/TP-AMI.2007.1005

Doddington G, Liggett W, Martin A, Przybocki M, Reynolds D (1998) Sheeps, goats, lambs and wolves: a statistical analysis of speaker performance in NIST 1998 speaker recognition evaluation. Proc ICSLD 13:1–5

Dunn SM, Keizer RL, Yu J (1989) Measuring the area and volume of the human body with structured light. IEEE Trans Syst Man Cybern 19(6):1350–1364. doi:10.1109/21.44059

Gokberk B, Dutagaci H, Ulas A, Akarun L, Sankur B (2008) Representation plurality and fusion for 3-D face recognition. IEEE Trans Syst Man Cybern Part B 38(1):155–173. doi:10.1109/TSMCB.2007.908865

Hartley R (2000) Multiple view geometry in computer vision. Cambridge University Press, Cambridge. ISBN 0521540518

Hernandez C, Vogiatzis G, Cipolla R (2008) Multiview photometric stereo. IEEE Trans Pattern Anal Mach Intell 30(3):548–554. doi:10.1109/TPAMI.2007.70820

Huang DS, Jia W, Zhang D (2008) Palmprint verification based on principal lines. Pattern Recogn 41(4):1316–1328. doi:10.1016/j.patcog.2007.08.0-16

Hu G, Stockman G (1989) 3-D surface solution using structured light and constraint propagation. IEEE Trans Pattern Anal Mach Intell 11(4):390–402. doi:10.1109/34.19035

Indovina M, Uludag U, Snelick R, Mink A, Jain A (2003) Multimodal biometric authentication methods: a COTS approach. In: Proceedings of MMUA 2003, Workshop multimodal user authentication, pp 99–106. http://www.nist.gov/customcf/ge-t_pdf.cfm?pub_id=151579

Kakadiaris IA, Passalis G, Toderici G, Murtuza MN, Lu YL, Karampatziakis N, Theoharis T (2007) Three-dimensional face recognition in the presence of facial expressions: an annotated deformable model approach. IEEE Trans Pattern Anal Mach Intell 29(4):640–649. doi:10.1-109/TPAMI.2007.1017

Kong AW, Zhang D (2004) Competitive coding scheme for palmprint verification. Proc Int Conf Pattern Recogn 1:520–523. doi:10.1109/ICPR.2004.1334184

Kong AW, Zhang D, Kamel M (2006) Analysis of brute-force break-Ins of a palmprint authentication system. IEEE Trans Syst Man Cybern Part B 36(5):1201–1205. doi:10.1109/TSMCB.2006.876168

Kühnel W (2006) Differential geometry: curves-surfaces-manifolds. American Mathematical Society, Providence. ISBN 0821839888

Saldner HO, Huntley JM (1997) Temporal phase unwrapping: application to surface profiling of discontinuous objects. Appl Opt 36(13):2770–2775. doi:10.1364/AO.36.002770

Samir C, Srivastava A, Daoudi M (2006) Three-dimensional face recognition using shapes of facial curves. IEEE Trans Pattern Anal Mach Intell 28(11):1858–1863. doi:10.1109/TPAMI.2006.235

Sanderson AC, Weiss LE, Nayar SK (1988) Structured highlight inspection of specular surfaces. IEEE Trans Pattern Anal Mach Intell 10(1):44–55. doi:10.1109/34.3866

Sansoni G, Biancardi L, Minoni U, Docchio F (1994) A novel, adaptive system for 3-D optical profilometry using a liquid crystal light projector. IEEE Trans Instrum Meas 43(4):558–566. doi:10.1109/19.310169

Snelick R, Uludag U, Indovina M, Jain A, Mink A (2005) Large-scale evaluation of multimodal biometric authentication using state-of-the-art systems. IEEE Trans Pattern Anal Mach Intell 27(3):450–455. doi:10.1109/TPAMI.2005.57

Srinivassan V, Liu HC (1984) Automated phase measuring profilometry of 3D diffuse object. Appl Opt 23(18):3105–3108. doi:10.1364/AO.23.003105

Stockman GC, Chen SW, Hu G, Shrikhande N (1988) Sensing and recognition of rigid objects using structured light. IEEE Control Syst Mag 8(3):14–22. doi:10.1109/37.472

Sun ZN, Tan TN, Wang YH, Li SZ (2005) Ordinal palmprint representation for personal identification. Proc IEEE Int Conf Comput Vis Pattern Recogn 1:279–284. doi:10.1109/CVPR.2005.267

Vapnik V (1995) The nature of statistical learning theory. Springer, Berlin. ISBN 0387987800

Woodard DL, Flynn PJ (2005) Finger surface as a biometric identifier. J Comput Vis Image Underst (CVIU) 100(3):357–384. doi:10.1016/j.cviu.2005.06.003

Yan P, Bowyer KW (2007) Biometric recognition using 3D ear shape. IEEE Trans Pattern Anal Mach Intell 29(8):1297–1308. doi:10.1109/TPAMI.2007.1067

Zhang D (2004) Palmprint authentication. Kluwer Academic Publishers, Dordrecht. ISBN 1-4020-8096-4

Zhang D, Shu W (1999) Two novel characteristics in palmprint verification: datum point invariance and line feature matching. Pattern Recogn 32:691–702. doi:10.1016/S0031-3203(98)00117-4

Chapter 8
3D Palmprint Classification by Global Features

Abstract Three-dimensional palmprint has proved to be a significant biometrics for personal authentication. Three-dimensional palmprints are harder to counterfeit than two-dimensional palmprints and more robust to variations in illumination and serious scrabbling on the palm surface. Previous work on three-dimensional palmprint recognition has concentrated on local features such as texture and lines. In this chapter, we propose three novel global features of three-dimensional palmprints which describe shape information and can be used for coarse matching and indexing to improve the efficiency of palmprint recognition, especially in very large databases. The three proposed shape features are Maximum Depth of palm center, Horizontal Cross-section Area of different levels and Radial Line Length from the centroid to the boundary of three-dimensional palmprint horizontal cross-section of different levels. We treat these features as a column vector and use Orthogonal Linear Discriminant Analysis to reduce their dimensionality. We then adopt two schemes: (1) coarse-level matching and (2) Ranking Support Vector Machine to improve the efficiency of palmprint recognition. We conducted a series of three-dimensional palmprint recognition experiments using an established three-dimensional palmprint database and the results demonstrate that the proposed method can greatly reduce penetration rates.

Keywords 3D palmprint identification • Global features • Palmprint indexing • OLDA • Ranking SVM

8.1 Introduction

Palmprint recognition has now been a topic of research for over ten years. Like other biometrics, palmprints demonstrate the properties required for personal authentication: universality, uniqueness, permanence, collectability and acceptability (Bolle et al. 2003). Furthermore, palmprints have some advantages over other

D. Zhang and G. Lu, *3D Biometrics*, DOI: 10.1007/978-1-4614-7400-5_8,   135
© Springer Science+Business Media New York 2013

biometrics. Palmprints are larger than fingerprints and therefore more robust to scars and dirt. Palmprint images are cheaper to collect and more acceptable than iris. Palmprints can distinguish between individuals more accurately than face and can also identify monozygotic twins (Kong et al. 2006).

Traditionally, palmprint recognition has made use of either high or low resolution 2D palmprint images. High resolution images are suitable for forensic applications (Jain and Feng 2009) while low resolution images are suitable for civil and commercial applications (Zhang et al. 2003). Most current research use low resolution palmprint recognition and is either texture-based or line-based. The texture-based methods include PalmCode (Zhang et al. 2003), Competitive Code (Kong and Zhang 2004) and Ordinal Code (Sun et al. 2005). These methods use a group of filters to enhance and extract the phase or directional features which can represent the texture of the palmprint. Line-based methods use line or edge detectors to explicitly extract line information from the palmprint that is then used for matching. The representative methods include Derivative of Gaussian based line extraction (Wu et al. 2006) and Modified Finite Radon transform (MFRAT) based line extraction (Huang et al. 2008).

In recent years, 3D techniques have been applied to biometrics authentication, such as 3D face (Samir et al. 2006) and 3D ear recognition (Yan and Bowyer 2007). Most recently, a structured-light imaging (Srinivassan and Liu 1984; Saldner and Huntley 1997) 3D palmprint system (Zhang et al. 2009) was developed that captures the depth information of a palmprint. This information is then used to calculate the Mean and the Gaussian curvatures for use in 3D palmprint matching and recognition. To date, however, there has been no work with 3D palmprints that has extracted global shape features, which may be useful in classification and indexing. For fingerprint, according to the global ridge structure and singularities, it can be classified into five classes: arch, tented arch, left loop, right loop and whorl (Henry 1900). Wu et al. (2004) classified the palmprint into six classes according to the palmprint principal lines. Besides the exclusive classification technique, the continuous classification technique is also widely used for indexing the database for personal identification (Lumini et al. 1997).

In this chapter, we propose extracting three novel global features from a 3D palmprint image: Maximum Depth (MD) at the center of the palm, Horizontal Cross-section Area (HCA) at different levels of the palm; and Radial Line Length (RLL) measured from the centroid to the boundary of the 3D palmprint. These features are then used to describe and classify the shape of the 3D palmprint using continuous classification. This involves first reducing the dimensionality of the features by treating these features as a column vector and applying Orthogonal Linear Discriminant Analysis (OLDA) (Ye 2005). We then improve the efficiency of palmprint recognition by indexing the database using coarse-level matching and Ranking Support Vector Machine (RSVM) (Joachims 2002).

The rest of the chapter is organized as follows. Section 8.2 describes how we define a region of interest for the 3D palmprint image and then extract our three proposed global features. Section 8.3 describes how global features can be used in classification in order to speed up identification. Section 8.4 gives the experimental results and Sect. 8.5 concludes the paper.

8.2 Global Features: Definitions and Extraction

The following describes our procedure for first extracting a region of interest from the 3D palmprint and then from that extract our proposed three global features.

8.2.1 The Region of Interest

Our definition and extraction procedure makes use of a 3D palmprint image containing 768×576 points captured using a structured-light imaging based 3D palmprint acquisition device (Zhang et al. 2009). First, we remove redundant and noisy boundary regions using a very simple Region of Interest (ROI) extraction process (Fig. 6.14). We segment a 400×400 points square that is respectively 68, 108, 234 and 134 points from the top, bottom, left and right boundaries of the 3D palmprint image as shown in Fig. 6.14a. Figure 6.14b shows the extracted ROI. After downsampling the 3D ROI to 200×200 points, we store it in a 200 by 200 matrix, $\{d_{ij} \mid i = 1, 2, \ldots, 200; \quad j = 1, 2, \ldots, 200.\}$, where d_{ij} is the depth value of the ith row and jth column point of the 3D ROI.

Our proposed 3D palmprint ROI extraction approach is much simpler than the one reported in (Zhang et al. 2003) and the extracted shape features are not sensitive to translation and rotation, which is why we can use such a coarse ROI extraction. As the shape feature is a form of global feature, we extract as large an ROI as possible. Of course, such a large ROI may contain noisy data so we use a mask to remove the noisy data according to the gradient of the 3D data. If the gradient of the point, which is defined as $|\nabla d| = \sqrt{\left(\frac{\partial d}{\partial x}\right)^2 + \left(\frac{\partial d}{\partial y}\right)^2}$, is larger than a given threshold, the point is regarded as noisy data.

Figure 8.1 shows a 3D ROI which contains noisy data and its corresponding mask. We use a 200 by 200 matrix, $\{m_{ij} \mid i = 1, 2, \ldots, 200; \quad j = 1, 2, \ldots, 200.\}$, to represent the mask, where $m_{ij} = 0$ is noisy data and $m_{ij} = 1$ is for other data.

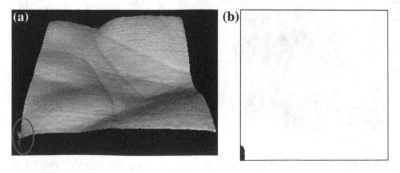

Fig. 8.1 a 3D ROI with noise. **b** Mask of the 3D ROI

8.2.2 Three Global Features

Using the ROI obtained from the original 3D palmprint data, we extract three kinds of features to describe the shape of the 3D palmprint: Maximum Depth (MD) of palm center, the Horizontal Cross-section Area (HCA) of different levels and the Radial Line Length (RLL) from the centroid to the boundary of 3D palmprint horizontal cross-section of different levels.

8.2.2.1 Maximum Depth

Maximum Depth (MD) means the maximum depth value of the 3D palm from a reference plane. The reference plane is decided using a rectangle as shown in the left of Fig. 8.2a. The depth of the reference plane d_r is the mean depth of the points contained by this rectangle.

$$d_r = \frac{1}{\sum\limits_{i=R_s}^{R_e}\sum\limits_{j=C_s}^{C_e} m_{ij}} \sum\limits_{i=R_s}^{R_e}\sum\limits_{j=C_s}^{C_e} (d_{ij} \cdot m_{ij}), \tag{8.1}$$

where d_{ij} is the depth value of the ith row and jth column point of the 3D ROI, m_{ij} is the corresponding mask value, R_s, R_e, C_s, and C_e respectively denote the start row, end row, start column and end column. The parameters $R_s = 65$, $R_e = 136$, $C_s = 6$ and $C_e = 35$ were set by experience. The reason we choose this region is that in the 3D ROI it appears to be relatively flat.

After getting the depth of the reference plane, we find the maximum depth, $d_{\max}$, in a region denoted by the right rectangle in Fig. 8.2a which starts at the

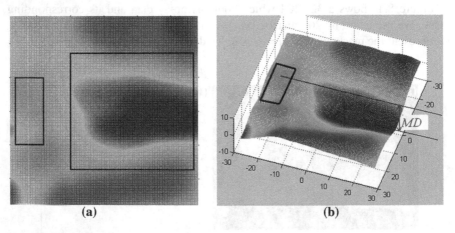

(a)   (b)

Fig. 8.2 Illustration of the MD feature (with color denoting the depth of the 3D ROI). **a** Location of the two *rectangles* used to calculate the reference plane and to find the maximum depth point. **b** Illustration of the MD feature

41st row and extends to the 160th row and from the 65th column to the 190th column. The MD can then be calculated easily by (2) as shown in Fig. 8.2b.

$$MD = d_{\max} - d_r. \tag{8.2}$$

8.2.2.2 Horizontal Cross-Section Area

To describe the shape of the 3D palmprint, we use a group of equidistant horizontal planes to cut the 3D ROI as shown in Fig. 6.17. Figure 8.3 shows a 3D ROI and its contour cut by the equidistant horizontal planes. To render the shape clearly, Fig. 8.3a only shows the 3D ROI image and hides the equidistant horizontal planes. In Fig. 8.3b, the blue curves denote deeper levels, the red curves denote higher levels and the remainder are medium levels. The Horizontal Cross-section Area (HCA) is defined as the area enclosed by the level curve. From Fig. 8.3b, we can see that most of the deeper level curves are enclosed and the areas are simply connected. These are more stable in response to noise or transformation.

To get a stable HCA, we take into consideration only the levels from the deepest point to the reference plane, defined in Sect. 8.3.1. Suppose we divide this region into N levels. Every level G^k, $k = 1, 2, \ldots, N$ is described with a 200×200 matrix and calculated by (3).

$$G_{ij}^k = \begin{cases} 1, & \text{if } d_{ij} > h \cdot (N - k + 1)/N \\ 0, & \text{others} \end{cases}, \ k = 1,2,\cdots,N; \ i = 1,2,\cdots,200; \ j = 1,2,\cdots,200. \tag{8.3}$$

where d_{ij} is the depth value of the ith row and jth column point of the 3D ROI and h is the palmprint depth defined by (2).

To make it more stable, we constrain every level growing from its previous level except the first level. That is

$$L^k = G^k \cap (L^{k-1} \oplus \Theta^{k-1}), \quad k = 2, 3, \ldots, N; \ L^1 = G^1. \tag{8.4}$$

Fig. 8.3 **a** A 3D ROI. **b** Its corresponding contour cutting by the equidistant *horizontal planes*

Where "∩" denotes logical AND, ⊕ denotes a morphological dilation operation and Θ^k is a disk morphological structuring element whose size can be calculated by $35 - 3 \times k$ (this is suitable for $N = 8$ by experience).

Figure 8.4 shows an example of all the levels stacked together. Figure 8.5 shows each of the levels separately.

After getting the cross-sectional levels L^k, $k = 1, 2, \dots, N$, the HCA, A^k, $k = 1, 2, \dots, N$ can be easily calculated by

$$A^k = \sum_{i=1}^{200} \sum_{j=1}^{200} L_{ij}^k. \tag{8.5}$$

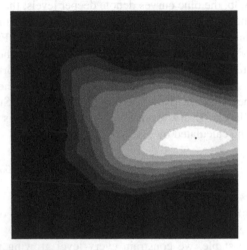

Fig. 8.4 An example of all the levels stacked together when $N = 8$

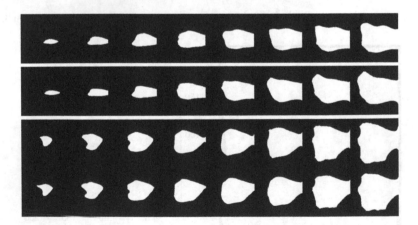

Fig. 8.5 The cross-sectional area feature (the *top* two *rows* are extracted from two samples collected from one palm; and the *bottom* two *rows* are extracted from two samples from another palm)

Fig. 8.6 *Radial line* starting from the centroid (*left* to *right*, $M = 8$, 16, 32 and 64 respectively)

8.2.2.3 Radial Line Length

The HCA is only a coarse description of the cross-section. To identify samples which have a similar cross-sectional area but have a different contour, we extract the Radial Line Length (RLL) feature which describes the shape of the contour. First, we calculate the centroid of the first level L^1, thereafter we treat it as the centroid of all levels. Then, from the centroid we draw M radial lines which intersect with the contour of every level. The distance between the intersection and the centroid is defined as the RLL. The radial lines are distributed at equal angles. We record these radial lines from the inner layers to the outer layers starting with the horizontal direction by an $M \times N$ dimensional vector, R_i, $i = 1, 2, \ldots, M \times N$, where M is the number of radial lines and N is the number of cross-sections. Figure 8.6 shows some examples of radial lines and their cross-sections. We can see that the RLL better represents the contour as the number of radial lines increases.

The above three global features are mainly determined by the central region of the palm. This region is certainly contained by the ROI described in Sect. 8.2.1 which makes these features insensitive to translation and rotation. Although the RLL feature can be affected by rotation as the contours change smoothly, if the rotation is small then the variation of the RLL feature will also be small. Actually, there are some restricting pegs on the capture device which can guide the user to put his/her hand on the proper place as described in (Zhang et al. 2009). Furthermore, we assume the user is cooperative when collecting data as we aim at civil rather than law enforcement applications.

8.3 Classification with Global Features

The classification of biometrics speeds up the identification process by reducing the number of comparisons that must be made. There are two kinds of classification techniques: exclusive classification and continuous classification. Both fingerprint (Henry 1900) and palmprint classifications (Wu et al. 2004) make use of exclusive classification. The main problem of this technique is that it uses only a small number of classes and the samples are unevenly distributed between them, with more than 90 % of the samples being in just two or three classes. A further problem with exclusive classification is that when classification is performed automatically, it is necessary to handle errors and rejected samples gracefully, which is a hard problem in practice. In contrast, for continuous classification, samples are not partitioned into disjoint classes but rather associated with numerical vectors which represent features of the samples. These feature vectors are extracted through a similarity-preserving transformation so that similar samples are mapped into close points in the multi-dimensional space (Maltoni et al. 2003). In this chapter, we adopt the continuous classification technique. As the global features combining MD, HCA and RLL are high-dimensional, we reduce the dimensions using the LDA method. We then improve the efficiency of palmprint recognition by applying coarse-level matching and Ranking Support Vector Machine (RSVM) to the low dimensional vectors.

8.3.1 Dimension Reduction Using Orthogonal LDA

LDA is a state-of-the-art dimensionality reduction technique widely used in classification problems. The objective is to find the optimal projection which simultaneously minimizes the within-class distance and maximizes the between-class distance, thus achieving maximum discrimination (Here, the "class" is used to denote the identity of the subjects, e.g. the samples collected from one palm are regarded as one class). However, the traditional LDA requires the within-class scatter matrix to be nonsingular, which means the sample size should be large enough compared with its dimension, but is not always possible. In this chapter, we therefore adopt the orthogonal LDA (OLDA) proposed in (Ye 2005), where the vectors of the optimal projection are calculated using the training database and the optimal projecting vectors are orthogonal to each other.

Suppose the 3D ROI has been divided to N levels and the M radial lines are used to represent the level contours. We can list the global features as a column vector, $F = \{MD, A^1, A^2, \ldots, A^N, R^1, R^2, \ldots, R^{N \times M}\}$, with $1 + N + N \times M$ rows. Given a training database which has n samples and k classes as $X = [X_1, X_2, \ldots, X_k]$, where $X_i \in \Re^{(1+N+N \times M) \times n_i}$, $i = 1, 2, \ldots, k$ and $n = \sum_{i=1}^{k} n_i$, adopting OLDA (Ye 2005) the optimal projection W can be calculated as follows.

First, the within-class scatter matrix S_w, the between-class scatter matrix S_b and total scatter matrix S_t can be expressed as

$$S_w = H_w H_w^T, \ S_b = H_b H_b^T, \ S_t = H_t H_t^T, \tag{8.6}$$

where

$$H_w = \frac{1}{\sqrt{n}} \left[X_1 - m_1 \cdot e_1^T, \ldots, X_k - m_k \cdot e_k^T \right], \tag{8.7}$$

$$H_b = \frac{1}{\sqrt{n}} \left[\sqrt{n_1}(m_1 - m), \ldots, \sqrt{n_k}(m_k - m) \right], \tag{8.8}$$

$$H_t = \frac{1}{\sqrt{n}} (X - m \cdot e^T), \tag{8.9}$$

where m_i is the centroid of the ith class X_i, m is the centroid of all the training samples X, $e_i = [1, 1, \ldots 1]^T \in \Re^{n_i}$, $i = 1, 2, \ldots, k$ and $e_i = [1, 1, \ldots 1]^T \in \Re^n$.

After calculating H_w, H_b and H_t, the reduced Singular Value Decomposition (SVD) is applied to H_t.

$$H_t \xrightarrow{\text{Reduced SVD}} U_r \Sigma_r V_r^T \tag{8.10}$$

Denote $B = \Sigma_r^{-1} U_r^T H_b$ and compute the SVD of B.

$$B \xrightarrow{\text{SVD}} U_B \Sigma_B V_B^T \tag{8.11}$$

Let

$$D = U_r \Sigma_r^{-1} U_B, \tag{8.12}$$

$$q = rank(B), \tag{8.13}$$

and denote D_q the first q columns of the matrix D. Then, compute the QR decomposition of D_q

$$D_q \xrightarrow{\text{QR decomposition}} QR, \tag{8.14}$$

where Q is the desired orthogonal matrix and optimal projection, i.e. $W = Q$.

After getting the optimal projection W, we can map the $1 + N + N \times M$ dimensional vector F to a lower dimensional space

$$\tilde{F} = W^T F, \tag{8.15}$$

where $\tilde{F} = \{f_1, f_2, \ldots, f_\Gamma\}$ is a Γ dimensional vector with $\Gamma < 1 + N + N \times M$.

8.3.2 Coarse-Level Matching

As the purpose of coarse-level matching is to speed up the identification during retrieval, we can regard it as a continuous classification approach. After mapping the global features' Γ dimensional vector $\tilde{F}$, the global features can be used to measure the similarity of two samples as follows:

$$\Delta = \left\| \tilde{F}^1 - \tilde{F}^2 \right\|_2 = \sum_{i=1}^{\Gamma} \left(f_i^1 - f_i^2 \right)^2 \cdot \left(f_i^1 - f_i^2 \right)^2. \tag{8.16}$$

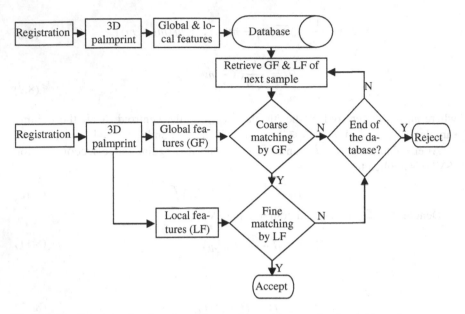

Fig. 8.7 The flowchart of registration and recognition with coarse-level matching scheme

In 3D palmprint identification, we can use the Γ dimensional global features to carry out coarse-level matching as shown in Fig. 8.7. If the testing sample passes coarse-level matching, it undergoes fine-level matching using 3D palmprint local features. If it does not pass, it moves on to the next sample in the database and so on until it has accessed the last sample in the database. From (8.16), we can see that coarse-level matching requires only Γ times of addition and multiplication which is much faster than fine-level matching using local features. Equation (8.17) gives the fine-level matching by Mean Curvature Image (MCI) feature (Zhang et al. 2009):

$$Y = \frac{2 \sum_{i=1}^{n} \sum_{j=1}^{m} Z_d(i,j) \cap Z_t(i,j)}{\sum_{i=1}^{n} \sum_{j=1}^{m} Z_d(i,j) + \sum_{i=1}^{n} \sum_{j=1}^{m} Z_t(i,j)}, \tag{8.17}$$

where symbol "$\cap$" represents the logical AND operation, Z_d and Z_t are the two binarized MCI features. To deal with the translation problem of ROI when calculating the matching score by (8.17), we will shift two, four, six and eight pixels of the test image along 8 directions: right, left, up, down, left-up, left-down, right-up and right-down, respectively. Adding the non-shift one, we will have $8 \times 4 + 1 = 33$ matching scores and the maximum one is selected. Suppose the size of the MCI feature is 128×128, i.e. $m = 128$ and $n = 128$, from (8.16) and (8.17) we can see that coarse-level matching is much faster than fine-level matching.

8.3.3 Ranking Support Vector Machine

Coarse-level matching scheme is a simple and easy way to reduce retrieval times. It's more useful for palmprint recognition if we can rank the candidate samples in the database in descending order according to the above global features. Searching for the closest matches to a given query vector in a large database is time consuming if the vector is even moderately high-dimensional. Various methods have been proposed to speed up the nearest neighbor retrieval, including hashing and tree structures (Matei et al. 2006). However, the complexity of these methods grows exponentially with increasing dimensionality (Chen and Bhanu 2009). Therefore, we have adopted the Ranking Support Vector Machine (RSVM) method (Joachims 2002), inspired by the approaches of internet search engines, to rank the candidate samples in the database.

Given a query q_k, $k = 1, 2, \ldots, n$ and a sample collection $D = \{d_1, d_2, \ldots, d_m\}$, the optimal retrieval system should return a ranking r_k^*, $k = 1, 2, \ldots, n$ that orders the samples in D according to their relevance to the query. In this chapter, the query q and the sample d are the Γ dimensional global features as described above. In our approach, if a sample d_i is ranked higher than d_j in some ordering r, i.e. $r(d_i) > r(d_j)$, then $(d_i, d_j) \in r$, otherwise $(d_i, d_j) \notin r$. Consider the class of linear ranking functions

$$(d_i, d_j) \in f_{\vec{w}}(q) \Leftrightarrow \vec{w}\phi(q, d_i) > \vec{w}\phi(q, d_j), \tag{8.18}$$

where $\vec{w}$ is a weighted vector that is adjusted by learning and $\phi(q, d)$ is a pairwise distance function describing the match between q and d can be defined as $\phi_i(q, d) = |q_i - d_i|, i = 1, \ldots, \Gamma$. Our goal is to find the optimal ranking function that will satisfy the maximum number of the following inequalities.

$$\forall (d_i, d_j) \in r_1^* : \vec{w}\phi(q_1, d_i) > \vec{w}\phi(q_1, d_j)$$
$$\cdots \tag{8.19}$$
$$\forall (d_i, d_j) \in r_n^* : \vec{w}\phi(q_n, d_i) > \vec{w}\phi(q_n, d_j).$$

It is easier to solve this problem if it is converted into to the following SVM classification problem by introducing non-negative slack variable $\xi_{i,j,k}$.

Hence, the problem is to minimize:

$$V(\vec{w}, \vec{\xi}) = \tfrac{1}{2}\vec{w} \cdot \vec{w} + C \sum \xi_{i,j,k}, \tag{8.20}$$

subject to:

$$\forall (d_i, d_j) \in r_1^* : \vec{w}\left(\phi(q_1, d_i) - \phi(q_1, d_j)\right) \geq 1 - \xi_{i,j,1}$$
$$\cdots \tag{8.21}$$
$$\forall (d_i, d_j) \in r_n^* : \vec{w}\left(\phi(q_n, d_i) - \phi(q_n, d_j)\right) \geq 1 - \xi_{i,j,n},$$

$$\forall i \forall j \forall k : \xi_{i,j,k} \geq 0, \tag{8.22}$$

where C is a parameter that allows trading-off margin size against training error and $C = 0.1$ set by experience.

In the training stage, it is the inner-class samples of a test sample that should be ranked higher than the inter-class samples, e.g. inner-class samples rank is 1 and inter-class samples rank is 0. We input the ranks together with the Γ dimensional global features into the RSVM algorithm to learn the optimal ranking function $f_{\vec{w}*}$. Given a new query q, the samples in the database can be sorted by their value of

$$rsv(q, d_i) = \vec{w}^* \phi(q, d_i). \tag{8.23}$$

8.4 Experimental Results

We used the 3D palmprint acquisition device developed in (Zhang et al. 2009) to establish a 3D palmprint database containing 8,000 samples collected from 400 palms. The 3D palmprint samples were collected in two separated sessions, 10 samples in each session. The average time interval between the two sessions is one month. The collection procedure required volunteers to put their palms naturally and without force on the device. The original spatial resolution of the data was 768×576. After ROI extraction, the central part (400×400) was extracted and down-sampled to (200×200) for feature extraction and recognition.

The database was divided into a training part (the first session of 4000 samples) and a testing part (the second session of 4,000 samples). As described in Sect. 8.3, the dimension of the proposed global features is $1 + N + N \times M$. To select the value of M and N we carried out a series of verifications on the training database where the class of the input palmprint was known. Each of the 3D samples was matched with the remaining samples in the training database. A successful match is where the two samples are from the same class. This is referred to as intra-class matching and the candidate image is said to be genuine. An unsuccessful match is referred to as inter-class matching and the candidate image is said to be an impostor. Treating the global features as a point in the $1 + N + N \times M$ dimension space, we simply use the Euclidian distance as the matching score. Table 8.1 shows the Equal Error Rate (EER) for N = 4, 8, 16 and M = 8, 16, 32, 64. The best result is $N = 8$ and $M = 32$.

In order to balance accuracy and efficiency, we chose $N = 8$ and $M = 32$ in the following experiments. This means the global features have $1 + N + N \times M = 265$ dimensions. Table 8.2 shows the verification results by MD, HCA, RLL and their combined results. From the last column of Table 8.2 we can see using the combined three global features will achieve a lower EER than each of the individual features.

Table 8.1 The EER of 3D palmprint verification for $N = 4, 8, 16$ and $M = 8, 16, 32, 64$ by MD + HCA + RLL

	M = 8 EER (%)	M = 16 EER (%)	M = 32 EER (%)	M = 4 EER (%)
N = 4	14.30	19.15	14.35	14.07
N = 8	14.20	16.30	12.32	12.54
N = 16	18.11	18.35	15.21	14.11

Table 8.2   The EER of 3D palmprint verification for $N = 8$ and $M = 32$ by MD, HCA, RLL and their combined result

Global features	MD	HCA	RLL	MD + HCA + RLL
EER (%)	25.8	20.4	18.6	12.32

Table 8.3   3D palmprint recognition rate by OLDA for different dimensions

Γ	1	2	3	4	5	6	7	8	9	10	12	15	20	30
Recognition rate (%)	2.0	9.6	26.0	42.6	56.6	66.2	74.6	78.4	82.4	83.8	86.0	89.6	90.4	93.6

Fig. 8.8   The plot of 3D palmprint recognition rate by OLDA for different dimensions

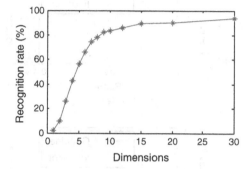

As described in Sect. 8.3.1, we use the OLDA method to reduce global features to a lower Γ dimension. To decide the optimal value of Γ, we carried out a series of recognition experiments on the 4,000 sample training database. We divided this database into two equal parts and then chose the first five samples of every palm for training and set aside the rest for testing. As shown in Sect. 8.3.1, Γ is equal to q in Eq. 8.13. Instead of $q = rank(B)$, we let $q = 1, 2, \ldots, 10, 12, 15, 20, 30$. Table 8.3 and Fig. 8.8 show the recognition results. We can see that 15 dimensions is a good choice for the following coarse-level matching and RSVM schemes.

Figure 8.9 shows the genuine and imposter distributions when the 3D palmprint 15-dimensional global features are applied to the 4,000 samples in the training database. Figure 8.9a–o are obtained by using the Euclidian distance to match the single dimension value from the 1st to 15th. Figure 8.9p shows the result of using the Euclidian distance to match all 15 dimensional values.

We next carried out the 3D palmprint classification and recognition experiments using the first sample of each class in the training database as a template and the 4,000 samples in the testing database as probes, making a total of 400 templates and 4,000 probes. The performance of classification and recognition is usually measured by error rate and penetration rate calculated in (Maltoni et al. 2003) as follows:

$$\text{error rate} = \frac{\text{number of false match}}{\text{total number of probe}} \times 100\,\%, \tag{8.24}$$

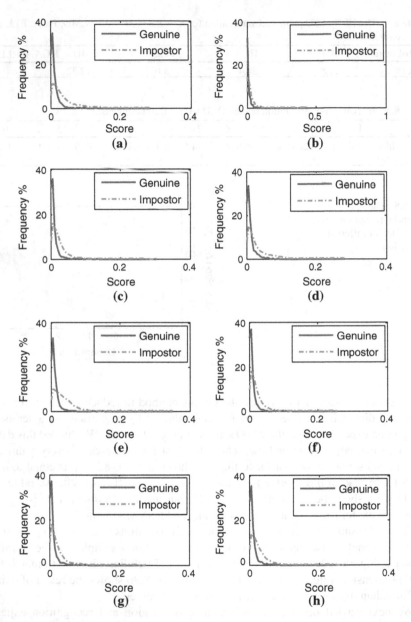

Fig. 8.9 Genuine and imposter distributions by the 3D palmprint 15-dimensional global features. **a–o** are obtained by matching the single dimension value from the 1st to 15th and **p** is obtained by matching all 15 dimensional values together

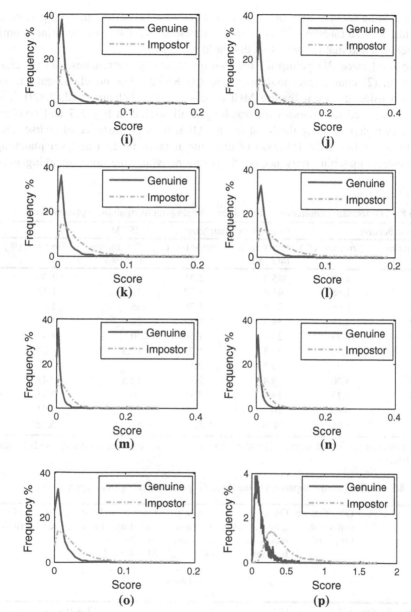

Fig. 8.9 (continued)

$$\text{penetration rate} = \frac{\text{number of accessed template}}{\text{total number of template in the database}} \times 100\%. \quad (8.25)$$

Obviously there is a trade-off between error rates and penetration rates. Generally speaking, if there is no classification, there are two retrieval strategies: (1) all of the templates in the database are visited and the template that gives the best matching score is regarded as the matched template, if the matching score is less than a given threshold Ψ_T; (2) given a threshold Ψ_t, the search continues until a match is found that is below that threshold.

We used three 3D palmprint recognition matching approaches: (1) no classification; (2) coarse-level matching; and (3) RSVM. For no classification, we matched using the local feature MCI as described in (Zhang et al. 2009). The process we used for coarse-level matching is illustrated in Fig. 8.7 and involves fine-level matching using the local feature MCI. A single instance of coarse-level matching requires only 1/36,000 of the time it takes to do fine-level matching (coarse-level matching only needs 15 operations while fine-level matching must

Table 8.4 Performance comparison of the three 3D palmprint recognition approaches

No classification		Coarse-level matching		RSVM	
Penetration rate (%)	Error rate (%)	Penetration rate (%)	Error rate (%)	Penetration rate (%)	Error rate (%)
100[a]	1.29	45.2	1.31	30	1.30
51.1	1.68	41.6	1.32	27.5	1.33
49.3	1.86	38.8	1.36	25	1.37
47.2	2.10	34.7	1.42	22.5	1.42
45.9	2.33	29.1	1.56	20	1.49
43.7	2.60	23.5	1.78	17.5	1.63
42.6	2.95	20.5	2.12	15	1.87
41.5	3.30	18.4	2.39	12.5	2.48
40.3	3.75	13.3	3.16	10	3.35
38.1	4.80	10.1	4.30	7.5	4.41
35.9	5.86	8.6	5.79	5	5.88

[a] This row used retrieval strategies (1) and the remaining rows used retrieval strategies (2) for no classification

Table 8.5 Running time comparison of the three 3D palmprint recognition approaches

	Once feature extraction time (ms)	Once dimensionality reduction time (ms)	Ranking or coarse matching time for all templates in database (ms)	Once matching time by MCI (ms)	Total running time for one probe testing (ms)
No classification	112	0	0	0.86	$112 + 0.86 * 400 * 1.0 = 456$
Coarse-level matching	136	0.1	0.5	0.86	$136 + 0.1 + 0.5 + 0.86 * 400 * 0.452 = 292.09$
RSVM	136	0.1	1.56	0.86	$136 + 0.1 + 1.56 + 0.86 * 400 * 0.30 = 240.86$

Fig. 8.10 Plot of the
penetration rate and error rate
of the three 3D palmprint
recognition approaches

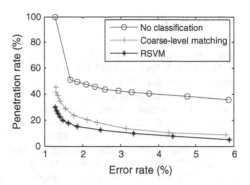

do $128 \times 128 \times (8 \times 4 + 1)$ operations, where 128×128 is the size of ROI and $8 \times 4 + 1$ is the shifting template times). For the above two approaches, the penetration rate and the error rate will vary with different thresholds Ψ_t. As for RSVM, we use the RSVM algorithm described in Sect. 8.3.3 to rank the templates in the database, and then match the top ρ percent by local feature MCI with the best matching score regarded as the matched template if this score is less than a given constant threshold Ψ_T. We can see from (25) that the ρ is equal to the penetration rate. Given different thresholds Ψ_t and ρ, we carried out a series of 3D palmprint recognition experiments. Tables 8.4, 8.5 and Fig. 8.10 show these experimental results. Even at an approximately equal error rate, the proposed coarse-level matching and RSVM approaches get a much lower penetration rate than the no classification approach. Obviously RSVM has the best performance but requires an additional offline training process compared to coarse-level matching.

8.5 Summary

This chapter proposed three global features for 3D palmprint images: Maximum Depth (MD), Horizontal Cross-section Area (HCA) and Radial Line Length (RLL). These cannot be extracted from 2D palmprints and are not correlated with local features, such as line and texture features. To make these global features efficient for use in coarse classification, we treat them as a multi-dimensional vector and use OLDA to map it to a lower dimensional space. We then improve the efficiency of 3D palmprint recognition using two proposed approaches, coarse-level matching and RSVM, both of which significantly reduce the penetration rate during retrieval. Our recognition experiments using an established 3D palmprint database of 8,000 samples show that the global features improve palmprint classification which greatly reduces search times.

References

Bolle RM, Connell JH, Pankanti S, Ratha NK, Senior AW (2003) Guide to biometrics. Springer, New York. ISBN 978-0-387-40089-1

Chen H, Bhanu B (2009) Efficient recognition of highly similar 3D objects in range images. IEEE Trans Pattern Anal Mach Intell 31(1):172–179. doi:10.1109/TPAMI.2008.176

Henry E (1900) Classification and uses of finger prints. Routledge, London

Huang DS, Jia W, Zhang D (2008) Palmprint verification based on principal lines. Pattern Recogn 41(4):1316–1328. doi:10.1016/j.patcog.2007.0-8.016

Jain AK, Feng JJ (2009) Latent palmprint matching. IEEE Trans Pattern Anal Mach Intell 31(6):1032–1047. doi:10.1109/TPAMI.2008.242

Joachims T (2002) Optimizing search engines using clickthrough data. In: Proceedings of the eighth ACM SIGKDD international conference on knowledge discovery and data mining, p 133-142, doi: 10.1145/775047.775067

Kong AW, Zhang D (2004) Competitive coding scheme for palmprint verification. Proc Int Conf Pattern Recogn 1:520–523. doi:10.1109/ICPR.2004.1334184

Kong AW, Zhang D, Lu GM (2006) A study of identical twins' palmprints for personal verification. Pattern Recogn 39(11):2149–2156. doi:10.10-16/j.patcog.2006.04.035

Lumini A, Maio D, Maltoni D (1997) Continuous versus exclusive classification for fingerprint retrieval. Pattern Recogn Lett 18(10):1027–1034. doi:10.1016/S0167-8655(97)00127-X

Maltoni D, Maio D, Jain AK, Prabhakar S (2003) Handbook of fingerprint recognition. Springer, U.S. ISBN 0-387-95431-7

Matei B, Shan Y, Sawhney H, Tan Y, Kumar R, Huber D, Hebert M (2006) Rapid object indexing using locality sensitive hashing and joint 3D signature space estimation. IEEE Trans Pattern Anal Mach Intell 28(7):1111–1126. doi:10.1109/TPAMI.2006.148

Saldner HO, Huntley JM (1997) Temporal phase unwrapping: application to surface profiling of discontinuous objects. Appl Opt 36(13):2770–2775. doi:10.1364/AO.36.002770

Samir C, Srivastava A, Daoudi M (2006) Three-dimensional face recognition using shapes of facial curves. IEEE Trans Pattern Anal Mach Intell 28(11):1858–1863. doi:10.1109/TPAMI.2006.235

Srinivassan V, Liu HC (1984) Automated phase measuring profilometry of 3D diffuse object. Appl Opt 23(18):3105–3108. doi:10.1016/0030-4018(92)90-606-R

Sun ZN, Tan TN, Wang YH, Li SZ (2005) Ordinal palmprint representation for personal identification. Proc IEEE Int Conf Comput Vis Pattern Recogn 1:279. doi:10.1109/CVPR.20-05.267

Wu XQ, Zhang D, Wang KQ, Huang B (2004) Palmprint classification using principal lines. Pattern Recogn 37(10):1987–1998. doi:10.1016/j.patco-g.2004.02.015

Wu XQ, Zhang D, Wang KQ (2006) Palm line extraction and matching for personal authentication. IEEE Trans Syst Man Cyber Part A 36(5):978–987. doi:10.1109/TSMCA.2006.871797

Yan P, Bowyer KW (2007) Biometric recognition using 3D ear shape. IEEE Trans Pattern Anal Mach Intell 29(8):1297–1308. doi:10.1109/TPAMI.2007.1067

Ye JP (2005) Characterization of a family of algorithms for generalized discriminant analysis on undersampled problems. J Mach Learn Res 6:483-502. http://jmlr.csail.mit.edu/papers/volume6/ye05a/ye05a.pdf

Zhang D, Kong AW, You J, Wong M (2003) On-line palmprint identification. IEEE Trans Pattern Anal Mach Intell 25(9):1041–1050. doi:10.1109/TPAMI.2003.1227981

Zhang D, Lu G, Li W, Zhang L, Luo N (2009) Palmprint recognition using 3-D information. IEEE Trans Syst Man Cybern Part C Appl Rev 39(5):505–519. doi:10.1109/TSMCC.2009.2020-790

Chapter 9
Joint Line and Orientation Features in 3D Palmprint

Abstract Two dimensional palmprint has been recognized as an effective biometrics identifier in the past decade. Recently, three-dimensional palmprint recognition was proposed to further improve the performance of palmprint systems. This chapter presents a simple yet efficient scheme for three-dimensional palmprint recognition. After calculating and enhancing the Mean: Curvature Image of the three dimensional palmprint data, we extract both line and orientation features from it. The two types of features are then fused at either score level or feature level for the final three-dimensional palmprint recognition. The experiments on the HKPU three-dimensional palmprint database which contains 8,000 samples from 400 palms show that the proposed feature extraction and fusion methods lead to promising performance.

Keywords 3D palmprint identification • Biometrics • Feature fusion • Mean curvature

9.1 Introduction

Automatic personal authentication using biometrics characteristics plays a key role in applications of public security, access control, forensics and e-banking, etc. Many kinds of biometrics authentication techniques have been developed based on different biometrics characteristics, such as fingerprint, face, iris, palmprint, hand shape, etc. Two dimensional (2D) palmprint recognition (Zhang et al. 2003; Kong and Zhang 2004; Sun et al. 2005; Ong et al. 2008; Li et al. 2006; Wu et al. 2006; Huang et al. 2008) has been widely studied in the past decade and it has been proven that palmprint is a unique biometrics identifier. 2D palmprint systems have merits of high accuracy and user friendliness, etc. Nonetheless, 2D palmprint can be easily counterfeited and much three dimensional (3D) palm structural information is lost. Inspired by the success of 3D techniques in biometrics

D. Zhang and G. Lu, *3D Biometrics*, DOI: 10.1007/978-1-4614-7400-5_9,
© Springer Science+Business Media New York 2013

authentication, such as 3D face (Samir et al. 2006) and 3D ear recognition (Yan and Bowyer 2007), most recently a structured-light imaging (Srinivassan and Liu 1984; Saldner and Huntley 1997) based 3D palmprint system (Zhang et al. 2009) was developed to capture the depth information of palmprint. In (Zhang et al. 2009), the Mean curvature and Gaussian curvature are calculated from the depth information and they serve as the basic features for 3D palmprint matching and recognition.

As shown in (Zhang et al. 2009), the Mean curvature is a stable and distinct feature of 3D palmprint. By normalizing and mapping the Mean curvature values to a plane, we can get a Mean Curvature Image (MCI) which contains line structure features and texture features of the 3D palmprint. In (Zhang et al. 2009), the MCI was binarized to highlight the line features and the binarized MCI was used as the feature map for 3D palmprint matching. However, the binarization operation loses much the texture information existing in the MCI. Actually, if we view the MCI of a 3D palmprint as a 2D palmprint image, then many 2D palmprint feature extraction techniques can be applied. In addition, the line and texture features could provide complementary information for palmprint discrimination. Therefore, in this chapter we propose to extract both line and texture features from MCI and fuse them efficiently for more accurate 3D palmprint recognition.

There are two main approaches to 2D palmprint recognition: line-based approach (Li et al. 2006; Wu et al. 2006; Huang et al. 2008) and texture-based approach (Kong and Zhang 2004; Sun et al. 2005). For the representative line-based methods, Li et al. proposed a modified line-based Hausdorff Distance for palmprint identification (Li et al. 2006); Wu et al. proposed a set of directional line detectors and used them to extract the palm lines for palmprint matching (Wu et al. 2006); Huang et al. proposed a modified finite Radon transform to extract the principal lines in palmprint (Huang et al. 2008). With respect to the texture-based methods, the most representative one may be Competitive Coding (CompCode) scheme proposed by Kong and Zhang (2004), where a series of directional Gabor: filters were used to extract the orientation features of palmprint. Sun et al. proposed an ordinal palmprint representation for personal identification (Sun et al. 2005).

With the MCI of 3D palmprint data, the line features can be easily extracted by setting a global threshold to segment the high curvature regions. For texture features, we can use six directional Gabor filters to extract the local orientation from the MCI, like what the CompCode method does on 2D palmprint images (Kong and Zhang 2004). In our preliminary work (Li et al. 2009), the line and orientation features are directly extracted from the MCI and they are fused at matching score level. In this chapter, we propose to use the Butterworth Low-Pass Filter (BLPF) to enhance the MCI before feature extraction, and fuse the extracted line and orientation features on the feature level. A series of experiments are conducted by using the HKPU 3D palmprint database, which contains 8,000 samples collected from 400 palms. The experimental results show that the proposed method is very promising, and significantly outperforms the methods in (Zhang et al. 2009) and (Li et al. 2009). Particularly, the proposed feature-level fusion can not only achieve higher accuracy than the score level fusion presented in (Li et al. 2009), but also require much less matching time.

The rest of the chapter is organized as follows. Section 9.2 discusses the calculation of MCI. Section 9.3 introduces the line and orientation feature extraction. Section 9.4 presents the matching and fusion scheme. Section 9.5 presents the experimental results and Sect. 9.6 concludes the chapter.

9.2 Mean Curvature Image Processing

9.2.1 Region of Interest Extraction

In (Zhang et al. 2009), we have developed a structured-light imaging based 3D palmprint acquisition device. The 3D palmprint data are typical range data which are represented by cloud points. Figure 9.1c shows a 3D palmprint sample (resolution: 768×576) collected by our device. For a better visualization of the 3D palmprint, Fig. 9.1c is rendered by OpenGL. We can see that in the 3D palmprint image, the cloud points in the boundary area and in the fingers are not suitable for feature extraction and recognition. Most of the useful and stable features locate in the center area of the palm. In addition, at different times when the user puts his/her hand on the collecting device, there will be some relative displacements of the positions of palm, even that we impose some constraints on the users to place their hands. Therefore, before feature extraction it is necessary to perform some preprocessing to align the palmprint and extract the central area of it, which is called the Region of Interest (ROI) extraction.

Our 3D palmprint data acquisition device can capture a 3D palmprint image and a 2D palmprint image simultaneously. As in (Zhang et al. 2009), we extract the 3D ROI with the aid of its corresponding 2D counterpart. Figure 9.1a shows a 2D palmprint image, the established local coordinate system by using the algorithm in (Zhang et al. 2003) and the ROI (i.e. the rectangle). Figure 9.1b shows

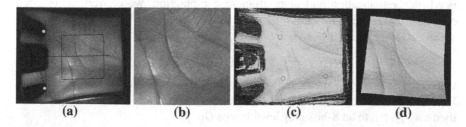

(a)   (b)   (c)   (d)

Fig. 9.1 The ROI extraction of 3D palmprint from its 2D counterpart. **a** The 2D palmprint image, the adaptively established coordinate system and the ROI (i.e. *the rectangle*), **b** The extracted 2D ROI, **c** The 3D palmprint image, whose cloud points have a one-to-one correspondence to the pixels in the 2D counterpart, **d** The obtained 3D ROI by extracting the cloud points corresponding to the pixels in 2D ROI

the extracted 256×256 2D ROI. Because the points in 3D palmprint image have a one to one correspondence to the points in its 2D counterpart, we can easily obtain the 3D ROI by extracting the cloud points corresponding to the 2D ROI. Figure 9.1d shows the extracted 3D ROI from Fig. 9.1c. (Note that we use a different viewpoint to show the 3D ROI in Fig. 9.1d.) By using the ROI extraction procedure, the 3D palmprint images are aligned so that the small translation and rotation introduced in the data acquisition process are corrected. In addition, the data amount used in the following feature extraction and feature matching process is significantly reduced. This will save much computational cost.

9.2.2 Curvature Calculation

With the ROI obtained from the original 3D palmprint data, stable and unique features are expected to be extracted for the following pattern matching and recognition. The Mean and Gaussian curvatures are intrinsic measures of a surface, i.e. they depend only on the surface shape but not on the way how the surface is placed in the 3D space (Kühnel 2003). Thus such curvature features are robust to the rotation, translation and even some deformation of the palm. From the work in (Zhang et al. 2009), we know that the Mean curvature is more informative than the Gaussian curvature in 3D palmprint recognition. Thus to save computation, we only consider the Mean curvature in the following development. We adopt the algorithm in (Besl and Jain 1988) to estimate the Mean curvature from 3D palmprint data for its simplicity and effectiveness:

$$H = \frac{(1 + (h_y)^2)h_{xx} - 2h_x h_y h_{xy} + (1 + (h_x)^2)h_{yy}}{2(1 + (h_x)^2 + (h_y)^2)^{3/2}}, \tag{9.1}$$

where h is the height of the points on the palmprint to the reference plane, h_x, h_y, h_{xx}, h_{yy} and h_{xy} are the first, second and hybrid partial derivatives of h to x and y coordinates separately.

With Eq. 9.1, the Mean curvatures of a 3D palmprint ROI can be calculated. For better visualization and more efficient computation, we convert the original curvature images into grey level images with integer pixels. We first normalize the Mean curvature value H to $\bar{H}$ as follows

$$\bar{H}(i,j) = 0.5(H(i,j) - \mu)/(4\delta) + 0.5, \tag{9.2}$$

where μ and δ are the mean and standard deviation of the curvature value. With Eq. 9.2, most of the curvature values will be normalized into the interval [0,1]. We then map $\bar{H}(i,j)$ to an 8-bits grey level image $G(i, j)$:

$$G(i,j) = \begin{cases} 0 & \bar{H}(i,j) \leq 0 \\ round\left(255 \times \bar{H}(i,j)\right) & 0 < \bar{H}(i,j) < 1 \\ 255 & \bar{H}(i,j) \geq 1 \end{cases} . \tag{9.3}$$

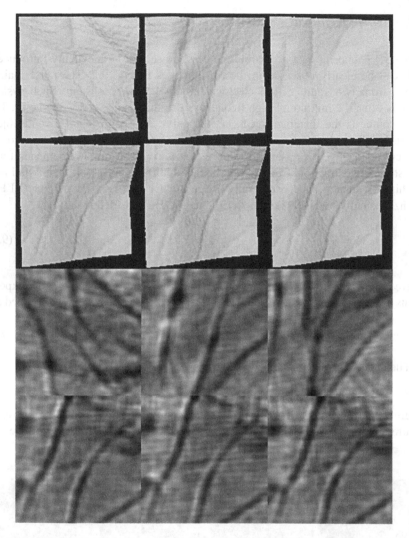

Fig. 9.2 The 3D ROI images (*first row*) and their MCI images (*second row*). The first three palmprint images are from different palms, while the last three palmprint images are from the same palm. (Please note that we change the viewpoint of 3D ROI for better visualization.)

We call image $G(i,j)$ the Mean Curvature Image (MCI). Figure 9.2 illustrates the MCI images from the same palm (at different times) and different palms. We can see that the 2D MCI images can well preserve the 3D palm surface features. Not only the principal lines, which are the most important and stable features in palmprint recognition, are clearly enhanced in MCI, but also the depth information of different shape structures is well preserved.

9.2.3 Noise Removal

In the 3D palmprint data acquisition process, noise will be inevitably introduced. This can be clearly observed in the MCI images in Fig. 9.2. The noise mainly comes from two sources. The electrical circuit hardware system. Such system noise is often the mixture of high frequency periodical noise and white noise. The other source is the imaging object, i.e. the palm. The palm is not a rigid object and its small deformations in the data collection process add random noise to the collected data. It is necessary to remove noise from the raw MCI images for a robust feature extraction. Considering the fact that these noise will mainly fall into the high frequency band, we simply use a Butterworth Low-Pass Filter (BLPF) (Gonzalez and Woods 2002) to reduce them. The BLPF is defined as:

$$H(u,v) = \frac{1}{1 + [D(u,v)/D_0]^{2n}}, \quad u = 1,2,\ldots,M, \quad v = 1,2,\ldots,N, \quad (9.4)$$

where $M * N$ is the image size; D_0 is the cut-off frequency (20 is used in our experiments); n is the order of BLPF and we set it to 4 by experience; $D(u, v)$ is defined as:

$$D(u,v) = \left[(u - M/2)^2 + (v - N/2)^2\right]^{1/2}. \quad (9.5)$$

The denoised MCI is obtained as follows:

$$G' = IFT(FT(G) \cdot H), \quad (9.6)$$

where FT denotes Fourier Transform, and denotes Inverse Fourier Transform. Figure 9.3 compares the MCI images before and after noise removal, and we can

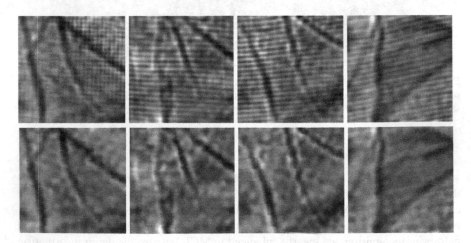

Fig. 9.3 Noise removal of MCI. The first row is the original MCI, and the second row is the MCI filtered by BLPF

clearly see that the image quality is much improved, which will benefit greatly the following feature extraction and matching.

9.3 Line and Orientation Feature Extraction

9.3.1 Line Feature Extraction

The principal lines and strong wrinkles are the most stable and significant features in the palmprint. In 3D palmprint, these features are represented by high curvature regions. So, it's very easy to extract the line feature from MCI by threshold:

$$L(i,j) = \begin{cases} 1 & G'(i,j) < c \cdot \mu_G \\ 0 & others \end{cases}, \tag{9.7}$$

where c is a constant and μ_G is the mean value of $G'(i,j)$. We set $c = 0.7$ in the experiments by experience. Note that binary image L can be directly used for matching. Figure 9.4 shows the binarized images of the MCI images in Fig. 9.3.

9.3.2 Orientation Feature Extraction

The line features extracted in Sect. 9.3.1 can indicate where the significant structures will happen in a palm, but the orientations of these line features are not implicitly represented. Apart from the line features, the MCI also has many finer texture features, which can be well characterized by local orientations as what has been done in 2D palmprint recognition (Kong and Zhang 2004). The Gabor filters have excellent capability to extract such features. By convolving the MCI with a series of Gabor filters along different orientations, the orientation along which the Gabor filter has the greatest response can be taken as the orientation of that point. The orientation features can then be coded and matched by angular distance for identification. This process is called the Competitive Coding scheme (Kong and

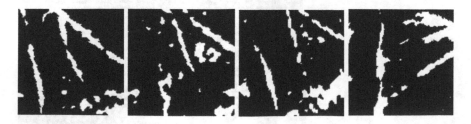

Fig. 9.4 The binarized MCI images. The white areas represent the high Mean curvature region position

Zhang 2004). In this chapter, the following Gabor filter is used for extracting the orientations (Lee 1996):

$$\psi(x, y, \omega, \theta) = \frac{\omega}{\sqrt{2\pi}\kappa} e^{-\frac{\omega^2}{8\kappa^2}(4x'^2 + y'^2)} \left(e^{i\omega x'} - e^{-\frac{\kappa^2}{2}} \right), \qquad (9.8)$$

where $x' = (x - x_0)\cos\theta + (y - y_0)\sin\theta$, $y' = -(x - x_0)\sin\theta + (y - y_0)\cos\theta$, and (x_0, y_0) is the center of the function; θ is the orientation of the Gabor functions in radians; $\omega = \kappa/\sigma$ is the radial frequency in radians per unit length. We set $\sigma = 4.2$ by experience, while κ is a coefficient defined by

$$\kappa = \sqrt{2\ln 2} \left(\frac{2^\alpha + 1}{2^\alpha - 1} \right), \qquad (9.9)$$

where α is the half-amplitude bandwidth of the frequency response. Here, we choose $\alpha = 1.3785$ octave by experience. More information about Gabor filters can be found in (Lee 1996). In this chapter we set the size of Gabor filter template as 35×35 with the center position $(17, 17)$.

Based on our experiments (please refer to Sect. 9.5 for details), we choose to use six Gabor filters with orientations $\theta = 0$, $\pi/6$, $2\pi/6$, $3\pi/6$, $4\pi/6$, $5\pi/6$ in the implementation for a good balance of accuracy and efficiency. Convolving the six filters with the MCI, and selecting the orientation which leads to the greatest response, we get the orientation features of MCI. Figure 9.5 shows an example, from which we can see that the extracted orientations can well represent the local directional structure in a neighborhood.

Fig. 9.5 The orientation map of an MCI

9.4 Line and Orientation Feature Matching and Fusion

9.4.1 Feature Matching

For the binary line feature map, we use the AND operation to calculate the matching score between two maps. Denote by L_d the binary MCI image in the database and by L_t the input MCI image. Suppose the image size is $n \times m$. The matching score between L_d and L_t is defined as:

$$R_L = \frac{2 \sum\limits_{i=1}^{n} \sum\limits_{j=1}^{m} L_d(i,j) \oplus L_t(i,j)}{\sum\limits_{i=1}^{n} \sum\limits_{j=1}^{m} L_d(i,j) + \sum\limits_{i=1}^{n} \sum\limits_{j=1}^{m} L_t(i,j)}, \tag{9.10}$$

where symbol "$\oplus$" means the AND logic operation. If L_d and L_t are identical, we will have the maximum matching score $R_L = 1$; on the contrary, if L_d and L_t are extremely different, the matching score will be $R_L = 0$.

For orientation features, we use integers $0 \sim 5$ to code the six orientations 0, $\pi/6$, $2\pi/6$, $3\pi/6$, $4\pi/6$, $5\pi/6$, respectively. Intuitively, the distance between parallel orientations should be 0, while the distance between perpendicular orientations should be 3. In other cases, the distance should be 1 or 2. Let D_d and D_t be the direction sets of the MCI images. The matching score between them can be defined as:

$$R_D = \frac{1}{3nm} \sum_{i=1}^{n} \sum_{j=1}^{m} F\left(D_d(i,j), D_t(i,j)\right), \tag{9.11}$$

where $F(\alpha, \beta)$ represents the angular distance between α and β

$$F(\alpha, \beta) = \min\left(|\alpha - \beta|, \ 6 - |\alpha - \beta|\right), \quad \alpha, \beta \in \{0,1,2,3,4,5\}. \tag{9.12}$$

Obviously, the value of $F(\alpha, \beta)$ can only be 0, 1, 2 or 3 as described above.

9.4.2 Fusion Scheme

9.4.2.1 Score Level Fusion

Suppose there are n matching scores and denote them by R_i, $i = 1, 2,...,n$. The commonly used score level fusion techniques include Min-Score (MIN) $R_{MIN} = \min(R_1, R_2, \cdots, R_n)$, Max-Score (MAX) $R_{MAX} = \max(R_1, R_2, \ldots, R_n)$, Summation (SUM) $R_{SUM} = \frac{1}{n} \sum\limits_{i=1}^{n} R_i$ and Weighted Average (WA) methods

(Snelick et al. 2005; Indovina et al. 2003). Because the Equal Error Rate (EER) is an important index of the matching result and it can be estimated by the training database, the weights can be determined according to the corresponding EER values. In (Snelick et al. 2005), a WA scheme, called Matcher Weighting (MW), is proposed:

$$R_{MW} = \sum_{i=1}^{n} w_i R_i, \quad w_i = \frac{1/e_i}{\sum_{j=1}^{n} 1/e_i}, \tag{9.13}$$

where w_i is the weight of R_i, and e_i is the corresponding EER. The MW scheme assigns smaller weights to those features with higher EER values. Here, we adopt this fusion scheme to fuse the matching scores obtained by line and orientation features.

9.4.2.2 Feature Level Fusion

For line and orientation features, after coding them to bit planes, it's very convenient to combine these bit planes for matching. The six types of orientation features can be coded to 3-bits planes as illustrated in Table 9.1 (Kong and Zhang 2004).

The binary line feature map can be readily represented by another bit plane (i.e. the 4th bit). Then, for each point in the MCI, we can use a 4-bits code to describe the line and orientation features. Let B_d and B_t be the 4-bits feature maps of two MCIs. The matching score between them can be efficiently calculated as follows:

$$R_f = \frac{\sum_{k=1}^{4} \sum_{i=1}^{n} \sum_{j=1}^{m} \left(\left(B_d^M(i,j) \cap B_t^M(i,j) \right) \cap \left(B_d^k(i,j) \otimes B_t^k(i,j) \right) \right)}{4 \sum_{i=1}^{n} \sum_{j=1}^{m} \left(B_d^M(i,j) \cap B_t^M(i,j) \right)}, \tag{9.14}$$

where B_d^M and B_t^M are two masks which indicate the non-palmprint pixels, B_d^k and B_t^k represent the kth bit plane of B_d and B_t, and $\otimes$ is the bitwise exclusive OR operation.

Table 9.1 Coding of orientation featurs

Orientation	Bit 1	Bit 2	Bit 3
0	0	0	0
1	0	0	1
2	0	1	1
3	1	1	1
4	1	1	0
5	1	0	0

9.5 Experimental Results

A 3D palmprint database has been established by using the 3D palmprint: imaging device developed by the Biometrics Research Centre, the Hong Kong Polytechnic University. The database is available at http://www4.comp.polyu.edu.hk/~biome trics/2D_3D_Palmprint.htm. The PolyU 3D palmprint database contains 8,000 samples from 200 volunteers, including 136 males and 64 females between 10 and 55 years old (Zhang et al. 2009). The 3D palmprint samples were collected in two separated sessions, and in each session 10 samples were collected from both the left and right hands of each subject. The average time interval between the two sessions is one month. The original spatial resolution of the data is 768×576. After ROI extraction, the central part (256×256) is used for feature extraction and recognition. The z-value resolution of the data is 32 bits. In data collection, each volunteer contributed samples from both his/her right-hand palm and left-hand palm. Samples collected from the same palm belong to the same class. Therefore, there are 400 classes and each class contains 20 samples in our database.

We performed two types of experiments on the established database: verification and identification. The experiments were performed under the Visual C++ 6.0 programming environment on a PC with Windows XP Professional operation system and Pentium 4 CPU of 2.66 GHz and 1 GB RAM. In verification, the class of the input palmprint is known and each of the 3D samples was matched with all the other 3D samples in the database. A successful matching is called intra-class matching or genuine if the two samples are from the same class. Otherwise, the unsuccessful matching is called inter-class matching or impostor. Using the established database, there are 31,996,000 matches in total.

In extracting the orientation features, the number of Gabor filters (refer to Sect. 9.3.2) should be determined. To this end, we performed a series of experiments by using different numbers of Gabor filters to extract the orientation features on the 3D palmprint database. The orientation θ of the Gabor filters is evenly partitioned in $[0, \pi)$. For example, if four Gabor filters are used, we have $\theta = 0, \pi/4, 2\pi/4, 3\pi/4$. θ Table 9.2 lists the equal error rate (EER) results by using 4 ~ 12 Gabor filters. We can see that the lowest EER is got by six Gabor filters. Using more than six Gabor filter cannot improve the accuracy but increase the computational cost. Therefore we use six Gabor filters ($\theta = 0, \pi/6, 2\pi/6, 3\pi/6, 4\pi/6, 5\pi/6$) in the following experiments.

Table 9.2 EER by using different numbers of Gabor filter to extract the orientation featurEs on 3D palmprint database

Number of Gabor filters	4	5	6	7	8	9	10	11	12
EER (%)	0.56	0.38	0.32	0.33	0.39	0.35	0.34	0.36	0.34

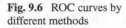

Fig. 9.6 ROC curves by
different methods

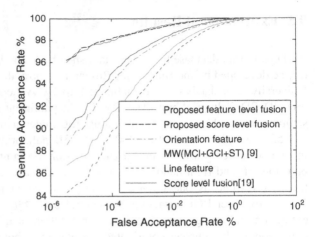

The verification experiments were performed by using each of the line and orientation features, as well as the fusion of them at score level and feature level respectively. We compared the proposed methods with the MW method in (Zhang et al. 2009), which fuses MCI, GCI and ST features on score level, and the score level fusion method in (Li et al. 2009). The ROC curves are shown in Fig. 9.6. The EER values are listed in Table 9.3, where the feature size, the preprocessing, feature extraction and matching time by using different features are also listed. The preprocessing in this chapter includes ROI extraction and BLPF based noise removal, while the preprocessing in (Zhang et al. 2009) and (Li et al. 2009) only includes ROI extraction. The BLPF filtering costs about 0.5 s but it improves the recognition accuracy significantly. From Fig. 9.6 and Table 9.3, we can see that the proposed fusion methods get much lower EER than other methods. Between score level fusion and feature level fusion, the later can achieve slightly better EER while requiring only half of the matching time. This implies that fusing the line and orientation features at feature level could be a more practical solution to real time 3D palmprint identification in a relatively large scale database.

The identification experiments were also conducted on the 3D palmprint database. In identification, we do not know the class of the input palmprint but want to identify which class it belongs to. In the experiments we let the first sample of each class in the database be template and use the other samples as probes. Therefore, there are 7,600 probes and 400 templates. The probes were matched with all the templates models, and for each probe, the matching results were ordered according to the matching scores. Then we can get the cumulative match curves as shown in Fig. 9.7. The cumulative matching performance, rank-one recognition rate and lowest rank of perfect recognition (i.e. the lowest rank when the recognition rate reaches 100 %) are listed in Table 9.4. From the experimental results we can see that the performance of the proposed fusion scheme is much better than the other three methods.

Table 9.3 Verification performance and running time by different methods

	Line feature		Orientation feature		Score level fusion		Feature level fusion		MW (Zhang et al. 2009)
With BLPF	Yes	No (Li et al. 2009)	Yes	No (Li et al. 2009)	Yes	No (Li et al. 2009)	Yes	No	No
Feature size (byte)	1568	1568	512	512	2080	2080	640	640	15680
EER (%)	0.53	0.688	0.32	0.495	0.17	0.284	0.16	0.268	0.45
Preprocessing time	1.116 s	660 ms	1.116 s	660 ms	1.116 s	660 ms	1.116 s	660 ms	660 ms
Feature extraction time (ms)	96	96	97	97	183	183	182	182	125
Matching time	0.35	0.35	0.15 ms	0.15	0.50	0.50	0.25	0.25	3.45

Fig. 9.7 CMC curves by
different methods

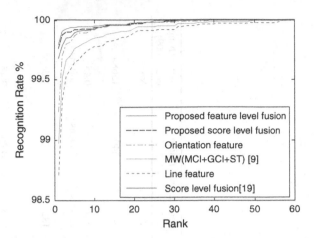

Table 9.4 Identification performance by different methods

	Line feature		Orientation feature		Score level fusion		Feature level fusion		MW (Zhang et al. 2009)
With BLPF	Yes	No (Li et al. 2009)	Yes	No (Li et al. 2009)	Yes	No (Li et al. 2009)	Yes	No	No
Rank-one recognition rate (%)	98.71	98.46	99.24	99.11	99.76	99.68	99.79	99.71	98.95
Lowest rank for perfect recognition	57	71	39	46	28	36	25	34	50

9.6 Summary

This chapter presented simple yet efficient schemes to extract and fuse the line
and orientation features of 3D palmprint for recognition. After the 3D palmprint
image was captured, the region of interest (ROI) was extracted, from which the
Mean Curvature Image (MCI) was calculated. By using the Butterworth Low-Pass
Filter (BLPF) to remove the high frequency noise, we extracted the line and orientation features from MCI, which are robust for palmprint recognition. The score
level and feature level fusion of the two types of features were proposed to match
and classify the palmprints. A series of verification and identification experiments
were performed on the HKPU 3D palmprint database with 8,000 samples from
200 individuals (400 palms). The experimental results showed that both the score
level and feature level fusion of line and orientation features can have much better

results than using only one of them and the existing 3D palmprint recognition methods. Particularly, the feature level fusion of line and orientation features can achieve better accuracy than score level fusion while requiring less matching time.

References

Besl PJ, Jain RC (1988) Segmentation through variable-order surface fitting. IEEE Trans PAMI 10(2):167–192. doi:10.1109/34.3881

Gonzalez RC, Woods RE (2002) Digital image processing, 2nd edn. Prentice Hall, Englewood Cliffs (ISBN: 0201180758)

Huang DS, Jia W, Zhang D (2008) Palmprint verification based on principal lines. Pattern Recogn 41(4):1316–1328. doi:10.1016/j.patcog.2007.08.016

Indovina M, Uludag U, Snelick R, Mink A, Jain A (2003) Multimodal biometric authentication methods: a COTS approach. In: Proceedings MMUA 2003, Workshop Multimodal User Authentication 99–106. http://www.nist.gov/customcf/ge-t_pdf.cfm?pub_id=151579

Kühnel W (2003) Differential geometry: curves-surfaces-manifolds, American Mathematical Society, ISBN: 0821826565

Kong AW, Zhang D (2004) Competitive coding scheme for palmprint verification. In: Proceedings of international conference on pattern recognition, vol 1, pp 520–523. doi: 10.1109/ICPR.2004.1334184

Lee TS (1996) Image representation using 2D Gabor wavelet. IEEE Trans PAMI 18(10):959–971. doi:10.1109/34.541406

Li F, Leung MKH, Yu XZ (2006) Palmprint matching using line features. In: International conference on advanced communication technology (ICACT 2006), pp 1577–1582. doi: 10.1109/ICACT.2006.206287

Li W, Zhang D, Zhang L (2009) Three dimensional palmprint recognition. In: IEEE international conference on systems, man, and cybernetics, pp 4847–4852. doi: 10.1109/ICSMC.2009.5346053

Ong MGK, Connie T, Jin TAB (2008) Touch-less palm print biometrics: novel design and implementation. Image Vis Comput 26(12):1551–1560. doi:10.1016/j.imavis.2008.06.010

Saldner HO, Huntley JM (1997) Temporal phase unwrapping: application to surface profiling of discontinuous objects. Appl Opt 36(13):2770–2775. doi:10.1364/AO.36.002770

Srinivassan V, Liu HC (1984) Automated phase measuring profilometry of 3D diffuse object. Appl Opt 23(18):3105–3108. doi:10.1364/AO.23.003105

Samir C, Srivastava A, Daoudi M (2006) Three-dimensional face recognition using shapes of facial curves. IEEE Trans PAMI 28(11):1858–1863. doi:10.1109/TPAMI.2006.235

Sun ZN, Tan TN, Wang YH, Li SZ (2005) Ordinal palmprint representation for personal identification. In: Proceeding of IEEE international conference on computer vision and pattern recognition, pp 279–284. doi: 10.1109/CVPR.20-05.267

Snelick R, Uludag U, Mink A, Indovina M, Jain A (2005) Large-scale evaluation of multimodal biometric authentication using state-of-the-art systems. IEEE Trans PAMI 27(3):450–455. doi:10.1109/TPAMI.2005.57

Wu XQ, Zhang D, Wang KQ (2006) Palm line extraction and matching for personal authentication. IEEE Trans Syst Man Cybern Part A Syst Hum 36(5):978–987. doi:10.1109/TSMCA.2006.871797

Yan P, Bowyer KW (2007) Biometric recognition using 3D ear shape. IEEE Trans PAMI 29(8):1297–1308. doi:10.1109/TPAMI.2007.1067

Zhang D, Kong AW, You J, Wong M (2003) On-line palmprint identification. IEEE Trans Pattern Anal Mach Intell 25(9):1041–1050. doi:10.1109/TPAMI.2003.1227981

Zhang D, Lu G, Li W, Zhang L, Luo N (2009) Palmprint recognition using 3-D information. IEEE Trans Syst Man Cybern Part C Appl Rev 39(5):505–519. doi:10.1109/TSMCC.2009.2020-790

Part IV
3D Fingerprint Identification by Multi-View Approach

Chapter 10
3D Fingerprint Acquisition Device

Abstract Touchless multi-view fingerprint technique provides a solution to capture 2D fingerprint image with larger effective area and 3D fingerprint image. This chapter thus presents a touchless multi-view 3D fingerprint capture system that acquires three different views of fingerprint images at the same time by using cameras placed on three sides. This device is designed by optimizing parameters regarding the captured fingerprint image quality and device size. A fingerprint mosaicking method is proposed to splice together the captured images of a finger to form a new image with larger useful print area. 3D finger shape is reconstructed through binocular stereo vision theory. Such 3D information is found very useful in fingerprint alignment. Experimental results show that the proposed mosaic method is more robust to low ridge-valley contrast fingerprint images than available methods. An EER of around 3.6 % by using SIFT features in the mosaicked images can be achieved while an EER of around 5.3 % is obtained by using only frontal images on our established database of 541 fingers, which shows the effectiveness of the proposed mosaic approach and our multi-view acquisition strategy. The efficiency of our device is further proved by comparing recognition accuracy between mosaicked images and touch based fingerprint images. Examples of fingerprint alignment by using the 3D information were for the first time shown, which proves contribution of 3D information to fingerprint recognition.

Keywords Touchless multi-view imaging • Scale invariant feature transformation (SIFT) • Mosaic • 3D fingerprint reconstruction

10.1 Introduction

Fingerprints have been used for personal identification for centuries and automatic fingerprint identification systems have been used for decades. Nowadays fingerprint technique has been widely used in both forensic and civilian applications. Compared with other biometrics features, fingerprint-based biometrics is the most

D. Zhang and G. Lu, *3D Biometrics*, DOI: 10.1007/978-1-4614-7400-5_10,   171
© Springer Science+Business Media New York 2013

proven technique and has the largest market shares. Although fingerprint recognition has been studied for many years and much progress has been made, even the performance of state-of-the-art matchers is still much lower than the expectations of people and theory estimation (Pankanti et al. 2002). Up to now, processing low quality latent prints still needs human intervention. In addition to the requirement for higher accuracy and speed, many new requirements are also raised along with more and more adoption of fingerprint technique in civilian applications, such as template security, hygiene and so on.

Fingerprint images can be acquired in off-line or on-line mode. The so-called ink-technique and extraction of latent fingerprints in crime scenes are examples of off-line acquisition. Nowadays on-line acquisition techniques have been widely used in most applications. Common on-line acquisition techniques include optical, solid-state, thermal and ultrasound (Xia and O'Gorman 2003). Optical capture devices may work in touch-based or touchless mode. Frustrated total internal reflection (FTIR) is a famous touch-based fingerprint imaging technique, which is used in most of government and forensic applications due to its excellent image quality. Touchless optical fingerprint imaging is actually not a novel technique. It uses cameras to directly image the fingertip. It has the advantages of hygiene and no latent prints. As its image quality is lower than that of FTIR and its size is bigger than that of solid-state sensors, this type of fingerprint devices is seldom found in the market. However, in recent years, with emergence of more applications, popularity of multimodal biometrics, and development of fingerprint algorithms, there is necessity of reconsidering touchless fingerprint imaging technique.

Earlier works about touchless fingerprint imaging device began from single-camera mode (Song et al. 2004; Mitsubishi Company 2012; Lumidigm Company 2012; TST Company 2012; Chen 2006). Song et al. (2004) designed a touchless fingerprint device using a monochrome CCD camera and double ring-type illuminators with blue LEDs. They stated that good quality images can be obtained by using the ring-type illuminators and some algorithmic amendments. Products of touchless fingerprint sensors from companies (e.g. Mitsubishi (Mitsubishi Company 2012), TST Biometrics (TST Company 2012) and Lumidigm (Lumidigm Company 2012) are on sale. Chen (2006) described a device that captures 3D shape of finger by using structured lights. Kumar and Zhou (2011) used a simple web camera to capture very low resolution fingerprint images. These kinds of devices all face the problem of view difference due to curvature of the finger shape. In real fingerprint recognition systems, the performance is degraded by the limited common area between fingerprints caused by view difference.

To deal with the above mentioned problem, multi-view touchless sensing techniques have been proposed (Fatehpuria et al. 2006; Parziale and Diaz-Santana 2006; TBS Company 2012; Choi et al. 2010). Typically, TBS (TBS Company 2012) proposed a 3D multi-camera touchless fingerprint device named Surround Imager™ by using five cameras to capture nail-to-nail fingerprint images at one time and provided the reconstructed 3D finger shape, as shown in Fig. 10.1 (Chen 2009).

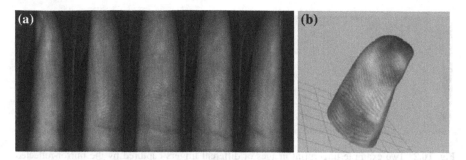

Fig. 10.1 Example images: **a** Different views of fingerprint images captured by Surround Imager™. **b** Reconstructed 3D finger shape

In the Chapter, they gave brief description to the device design and related algorithms about 3D reconstruction and recognition. However, the details of algorithms have not been given and performance evaluation has not been reported. Later, they continued to improve their device and developed new versions of products by using three cameras at one time (TBS Company 2012). The detail specifications of such devices and subsequent algorithms for image processing are not available. Choi et al. (2010) suggested using a single camera and two planar mirrors to form the multi-view fingerprint imaging device. The side views of the finger reflected by these mirrors are captured by the central camera to form multi-views of fingerprint images. This device has the advantage of less cost, but the hardware designing is very complex. For instance, the depth of field (DOF) of the central camera should be large enough to cover the three views of finger with high clarity. The setting of mirror and finger should be carefully considered due to the different size of finger. Such system also causes some difficulties to the post image processing. One is how to divide the whole image (as shown in Fig. 10.2) into three segments. Constant threshold is not suitable to different size of fingers, as Fig. 10.2a and b shown. Another one is about stereo calibration for 3D reconstruction. Current techniques for stereo calibration are mostly based on separate pictures captured by different cameras. The size of effective area of side-view images provided by mirror-reflected device is normally smaller than the one offered by multi-camera based device (see Figs. 10.1a and 10.2). We finally summarized the strengths and weaknesses of these two typical touchless multi-view fingerprint imaging systems in Table 10.1. The techniques used to process the captured fingerprint images are also listed in Table 10.1.

Due to the drawbacks and limited applications of mirror-reflected imaging technique, multi-camera mode is adopted in this Chapter to design our fingerprint capture device. Meanwhile, considering the drawbacks and difficulties to get detail specifications of existing multi-camera mode devices, as well as the unavailability of large scale touchless multi-view fingerprint databases in the public domain, This Chapter designed a touchless multi-view fingerprint capture device using multi-camera mode with optimized device parameters. Both image quality and device size are considered in designing the capture device. We established a database with

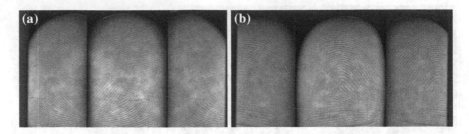

Fig. 10.2 Two example fingerprint images of different fingers captured by the mirror-reflected device

Table 10.1 Comparison of strengths, weaknesses and applications of two typical touchless multi-view fingerprint imaging systems

Device	Strengths	Weaknesses	Applications
Surround Imager™	Cover larger effective area; possible to achieve 3D reconstruction	Relatively expensive	3D fingerprint recognition. Unwrapping to get larger effective area
Mirror-reflected imaging device	Less cost	High hardware designing complexity; manually segment ROI; limited effective area	Mosaic to get larger effective area

541 fingers. Based on the established database, the applications of mosaic and 3D information are both studied. For mosaicking, we firstly extract the Scale Invariant Feature Transformation (SIFT) features on each image, followed by matching them using RANdom SAmple Consensus (RANSAC) algorithm to estimate the transformation model, then extract the stitching line by comparing the similarity of overlapping region, finally generate the mosaic image after Gaussian smoothing. The 3D finger shape is reconstructed given off-line camera calibration parameters, matched pairs between adjacent images in the mosaic step, and the estimated finger shape model. We found that such 3D information can be used to align two fingerprint images coarsely, which benefits fingerprint matching.

10.2 Acquisition Device Design

With the motivation of designing a cheaper and more optimized touchless multi-view fingerprint capture device, we studied and selected the system parameters in this section. The schematic diagram of the device is shown in Fig. 10.3. Cameras are focused at the finger. LEDs are used to light the finger and are arranged to give uniform brightness. A hole is designed to place the finger with fixed

Fig. 10.3 Schematic diagram of the proposed touchless multi-view fingerprint capture device

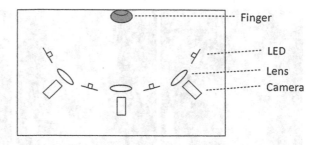

position. The main factors which influence the captured image quality, device size, and 3D reconstruction result mainly include the camera and lens configuration, distance between finger and lens, color of light source, and camera numbers and arrangement. The device proposed in this Chapter is designed based on fixed camera: JAI CV-A50. We then discussed the design of our device in detail as follows.

10.2.1 Lens Selection and Distance Setting

In order to capture fingerprint images with high quality and minimize the size of device, it is very important to select suitable lens and setting appropriate distance between object and lens. Because these two factors have impact on the captured image resolution, size of the effective fingerprint area and the height of the device. Different kinds of fingerprint features can be robustly extracted from different resolution images (Zhang et al. 2011). For traditional touch-based automated fingerprint identification systems, ~500 and ~800 dpi are required for minutiae and sweat pores, respectively (Zhang et al. 2011). For touchless-based systems, there is no survey showing which resolution is suitable. In (Fatehpuria et al. 2006), the resolution is larger than 500 dpi (700 dpi in center part and minimum of 500 dpi on image boarders). In Choi's paper (Choi et al. 2010), the resolution of captured image is ~500 dpi.

In this Chapter, we tried several kinds of resolutions to find an optimal one. Example fingerprint images at three kinds of resolution: ~750, ~500, and ~400 dpi are shown in Fig. 10.4. The corresponding lens focal length and object-to-lens distance is (25, 105 mm), (16, 145 mm), and (12, 91 mm) respectively. The image size is all restricted to 480 × 640 pixels. We finally set our device lens focal length and object-to-lens distance as (12, 91 mm) based on the following reasons. Firstly, we found that ridges on fingerprint images can be extracted at all of the above mentioned resolutions, as shown in Fig. 10.5. Secondly, the size of effective area is the largest one when resolution is at 400 dpi. Thirdly, the minimum object-to-lens distance is reached when resolution is ~400 dpi.

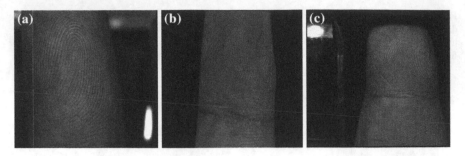

Fig. 10.4 Example fingerprint images with different resolutions: **a** ~750 dpi, **b** ~500 dpi, **c** ~400 dpi

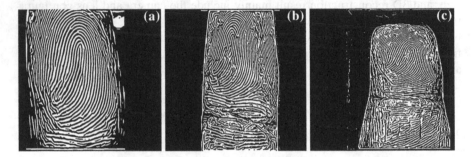

Fig. 10.5 Binarized fingerprint images with different resolutions: **a** ~750 dpi, **b** ~500 dpi, **c** ~400 dpi

10.2.2 Light Source Selection

Human skin has different luminous reflectance to different light sources (Elli 2001). Proper illuminator will help us to obtain touchless fingerprint images with high ridge-valley contrast. Among various kinds of light sources, blue LED and green LED are most popular ones.

In Parziale's paper (Parziale and Diaz-Santana 2006), authors demonstrated that green light provides a higher contrast than red and blue lights. In Wang's paper (Wang et al. 2009), authors studied how to get high-contrast contactless fingerprint images from aspects of polarization states, illumination wavelength, detection wavelength, and illumination direction. They offered systematic evidence that blue LED is the best choice among infrared LED, red LED, green LED and blue LED.

This Chapter thus captured several fingerprint images using blue LED and green LED as illuminator, binarized them using the same algorithm and parameters. Figure 10.6 shows an example images. We found that there is little difference between blue LED and green LED when binarizing the images by the same algorithm, as shown in Fig. 10.6c and f. The zoomed-in segment of binarized fingerprint images using blue LED is similar with the one using green LED. Indicated

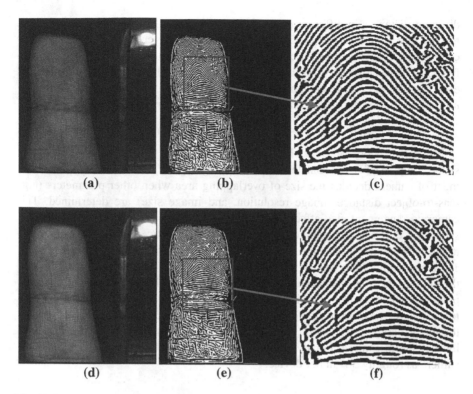

Fig. 10.6 Example fingerprint images using different light scources: **a** Original image captured by using blue LED. **b** Binarized image of (**a**). **c** Zoomed-in segment on (**b**). **d** Original image captured by using green LED. **e** Binarized image of (**d**). **f** Zoomed-in segment on (**e**)

by the strong evidence shown in Wang's paper (Wang et al. 2009), we finally chose blue LED as the light source in this Chapter.

10.2.3 Camera Numbers and Arrangement

The number of cameras directly decides the cost of the device. The smaller the number of cameras is, the cheaper the device will be. However, too few cameras cannot provide a panoramic view to the finger or lead to small overlapping region between side and frontal images. Given the value of resolution r and the size of the image $w * h$, we can easily calculate the size of measured area of the finger $w * h$ by Eq. (10.1). As mentioned above, when we set resolution as ~400 dpi and image size as 480×640 pixels in our device, the measured area will be 30.48×40.64 mm. It is bigger enough to cover the size of most fingers, which means the panoramic view of the finger can be captured by each camera in our device. However, the shape of human's fingers is curved, which leads to different distances in different parts of the finger. From Fig. 10.7, we can see that the distance from side parts to lens (i.e., D_2 or D_3) is larger than the distance from central part to lens (i.e., D_1). Perspective distortion is caused by these distance differences. To keep intact of each part (left, central, right) of the finger captured by each camera, three cameras, one central camera and two side cameras, are finally used in our device to capture different views of the finger, as shown in Fig. 10.3.

$$H = 25.4 \times h/r, W = 25.4 \times w/r \qquad (10.1)$$

Since the theory of binocular stereo vision is used to reconstruct 3D finger shape, only the 3D coordinates of overlapping region are available. The placement of camera decides the size of overlapping area when other parameters (e.g. lens-to-object distance, image resolution, and image size) are determined. The camera arrangement also affects the final mosaic image size. As illustrated in the schematic diagram of Surround Imager™, the angle between adjacent cameras is around 45°. While in the mirror-reflected device, they set the angle of mirrors as 15° empirically. Thus, in our design, we tried angles of 15°, 30°, 45°. Intuitively, the smaller the angle between adjacent cameras is, the larger the size of overlapping region is. However, the side view of fingers cannot be captured if the angle is too small. As the example images shown in Fig. 10.8. When the angle is set as 15°, the image captured by the left side camera is almost the same as the one captured by the central camera. Finally, we set the angle between central camera and side camera as roughly 30° in our device.

Fig. 10.7 Distance between lens and different parts of the finger

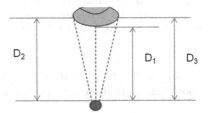

Fig. 10.8 Example fingerprint images captured by different cameras when the angle between adjacent camera is 15°: **a** Fingerprint image captured by the left camera. **b** Fingerprint image captured by the central camera

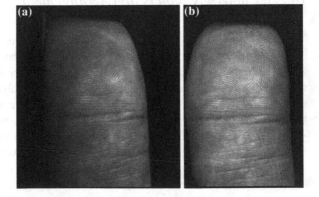

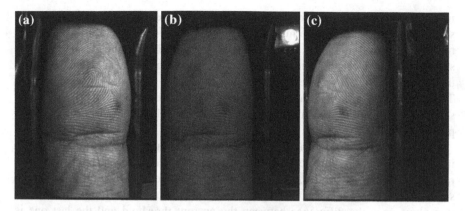

Fig. 10.9 Images of a finger captured by our device (*left, frontal, right*)

In conclusion, we designed the multi-view touchless fingerprint capture device shown in Fig. 10.3 with specific parameters mentioned above. The three view images of a finger captured by our device are shown in Fig. 10.9.

10.3 Device Applications

As shown in Fig. 10.9, we can get left-side, frontal, right-side fingerprint images at one time by our device, from which we can observe that ridge frequency increases from center to side and the contrast of ridge and valley is not high. To overcome the drawbacks of view difference and enlarge the size of effective area, one solution is to combine these three views of images into one. Low ridge-valley contrast is one intrinsic characteristic of touchless fingerprint imaging techniques, which is very hard to conquer from hardware. One way to alleviate this problem is to get more fingerprint features which are robust to the effect of low ridge-valley contrast from software. Fortunately, 3D coordinates of overlapping region provided by the device can be calculated using the theory of binocular stereo vision, which offers more information to assist touchless fingerprint recognition. This section studied two applications: mosaic and 3D reconstruction, based on the fingerprint images captured by our device.

10.3.1 Fingerprint Image Mosaic

Fingerprint image mosaicking is a technique for integrating different view of images into a continuous one with larger undistorted fingerprint area. The procedure of fingerprint image mosaicking mainly includes feature extraction, transform estimation, stitching line selection and post processing steps to generate smooth new image. In our proposed method, we firstly preprocess the original image, extract the scale invariant feature transformation (SIFT) feature point, and establish initial correspondences by point wise matching method. Then, the parameters of thin plate spline (TPS) model are estimated for aligning the side and frontal images. After that, the stitching line is selected from the overlapping region of adjacent images. The mosaicked fingerprint image is finally generated after smoothing. The overview schematic diagram of the algorithm is presented in Fig. 10.10 and details of the proposed approach are described as follows.

First, we should segment the image into foreground and background. The iterative thresholding segmentation method is adopted, which can easily separate the ROI region from the background. This method selects the threshold to segment the foreground and background region in iterative fashion. The iteration stops once the difference between the current threshold and the last one is smaller than 0.005 during the iterations. Finally ROI is extracted by the threshold. Figure 10.11b and e show the segmentation results of Fig. 10.11a and d.

Features frequently-used in fingerprint image mosaicking and matching contain minutiae, ridge map, and SIFT feature (Choi et al. 2010; Park et al. 2008; Feng 2008; Malathi and Meena 2010; Jain and Ross 2002; Shah et al. 2005; Choi et al. 2007). In this Chapter, we chose to use SIFT feature in our algorithms for the following reasons. Firstly, it is inevitable to capture fingerprint images with very low ridge-valley contrast by touchless imaging techniques. In this situation, minutiae and ridge features are hard to correctly extract, as the example shown in Fig. 10.12. Secondly, there are errors introduced when extracting minutiae or thinning ridges, which means it cannot reach at pixel accuracy. Thirdly, SIFT feature is robust to low image quality and deformation variation (Park et al. 2008). It also describes the local texture features exactly in pixel level and is rich in quantity (Malathi and Meena 2010).

Scale Invariant Feature Transformation (SIFT) (Lowe 2004) was popular in object recognition and image retrieval. It provides feature which is invariant to scale, rotation and affine transforms. There are four main steps to extract SIFT features. (1) The scale-space extrema is detected from images generated by applying multi-scales of difference of Gaussian (DoG) functions to the input image; (2) The accurate location of keypoint is determined according to the measurement of their

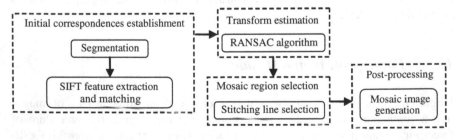

Fig. 10.10 The overall schematic diagram of the proposed fingerprint mosaicking method

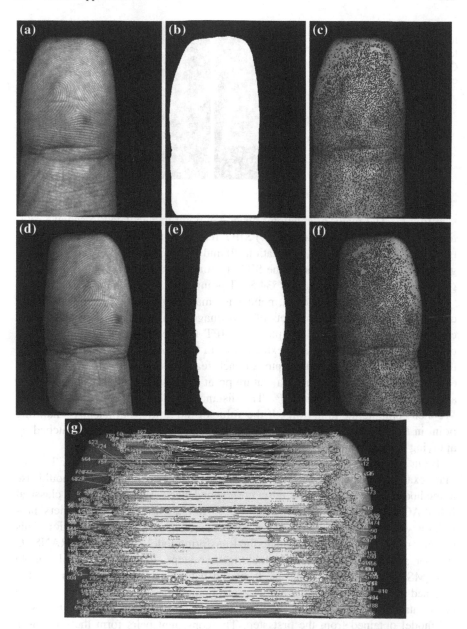

Fig. 10.11 Initial correspondences Establishment: **a** Original frontal image. **b** Segmentation result of (**a**) by iterative thesholding method. **c** Extracted SIFT feature from (**a**). **d** Original left-side image. **e** Segmentation result of (**d**) by iterative thesholding method. **f** Extracted SIFT feature from (**d**). **g** Initial correspondences establishment by point wise matching

Fig. 10.12 Example
fingerprint image with very
low ridge-valley contrast: **a**
Original image. **b** Ridge map.
c Extracted minutiae

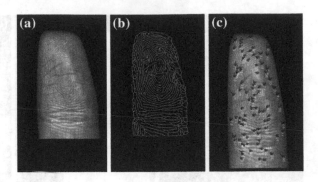

stability; (3) The major orientation of each keypoint is calculated to achieve rotation-invariant keypoint descriptor; (4) SIFT feature with four properties, i.e., spatial location (x, y), scale (s), orientation (θ) and keypoint descriptor (kd), is finally generated. Figure 10.11c shows the SIFT features extracted from an example fingerprint image (there are totally 7534 SIFT points).

After SIFT feature extraction, point wise matching method is adopted to find correspondences from feature sets of two images. This method is performed by comparing the associated descriptors of SIFT features. More specifically, given two SIFT feature sets P_1 and P_2 extracted from two images I_1 and I_2, we calculate the inner product between descriptor of each feature point in P_2 and descriptor of each feature point in P_1. For each feature point in P_1, we can find its closest point and the second closest point in P_2. The distance of them is labeled as d_1 and d_2. We then compute the ratio d_1/d_2. If the value of the ratio is sufficiently small, the point in P_1 is considered to match with the point in P_2. 811 pairs are matched by applying this method to Fig. 10.11a and d, as shown in Fig. 10.11g.

From Fig. 10.11g, we can see that there exist false correspondences. To estimate exact parameters of transform model between two images, we should use a method which is robust to false correspondences. Fortunately, the classical RANSAC algorithm which is insensitive to initial alignment and outliers provides a way to calculate the optimal model parameters in an iterative fashion. This Chapter used this method to estimate the transform model. The classical RANSAC (Fishler and Bolles 1981) mainly includes two steps. First, the minimal sample sets (MSSs) are randomly chosen from the dataset, and the parameters of the assumed global transformation model are estimated based on MSSs. Second, the other data in the dataset are checked to determine whether they are consistent with the model obtained from the first step. The consistent pairs form the consensus set (CS). RANSAC terminates when the probability of finding a better ranked CS drops below a certain threshold. Finally, the optimal transform model parameters and CS are both provided. It is notable that the model used in this method depends on the deformation form of the matched images. Due to the curved surface of finger and distortions introduced by cameras, we chose TPS model in the RANSAC algorithm. This model is popularly used in fingerprint domain (Choi et al. 2010, 2007; Ross et al. 2005). Figure 10.13 gives the CS when RANSAC with TPS

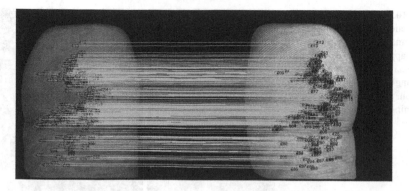

Fig. 10.13 CS of Fig. 10.11g calculated by model parameters obtained after RANSAC

model acted on the initial correspondences of Fig. 10.11g, which demonstrated the effectiveness of the algorithm.

Once the transform model parameters are obtained, we should determine how to stitch them to generate the final mosaic image. The approach we proposed consists of two stages. In the first phase, we extract the overlapping region of the two images. The width of the overlapping region is constrained by the maximal and minimal column coordinates given in the transformation estimation step. As shown in Fig. 10.14, the overlapping region on the frontal and left side images is framed by blue lines. In the second phase, we partition the overlapping region into sub-blocks with size of 21 × 21 pixels, and then calculate the correlation between the sub-block in left side image and the corresponding sub-block in frontal image. The location of the stitching line is defined as the center line of the sub-block which offered the largest correlation value. The red line in Fig. 10.14 shows the final stitching line we obtained by using our proposed method.

Due to the intensity difference of images captured by separate cameras, normalization is necessary to make the mosaicked image smooth. Here, we used the MAX_MIN strategy to all of the images based on the intensities of their overlapping regions. Then a Gaussian smoothing is applied to the mosaicked image to get the final result.

10.3.2 3D Finger Shape Reconstruction

The 3D information of an object can be obtained from its two different plane pictures captured at one time according to binocular stereo vision theory in computer vision. Our touchless multi-view fingerprint imaging device has been designed to provide three images of the finger simultaneously, which means 3D finger shape feature is available based on stereo vision and photogrammetry methods. There are mainly four steps for 3D fingerprint reconstruction, namely camera

Fig. 10.14 Stitching line
extraction: **a** Original left-
side image with rectangled
overlapping region. **b**
Original frontal image with
rectangled overlapping
region. **c** Final selected
stitching line

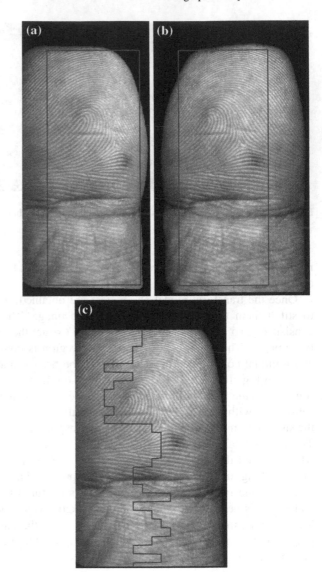

parameters calculation, correspondences establishment, 3D coordinate computation, and shape model estimation. In the first step, intrinsic parameters (Focal
Length, Principal Point, Skew, and Distortion) of each camera and extrinsic
parameters (Rotation, Translation) between cameras should be given (Sonka et al.
1999; Hartley and Zisserman 2003). It is called camera calibration in computer
vision and usually implemented off-line. We adopted the methodology proposed
in Zhang's paper (Zhang 2000) and the improved algorithm coded by Bouguet
(2010) for camera parameter calculation. It is notable that the position of the

middle camera is chosen as the reference system for the reason that central part of the fingerprint is more likely to be captured by the middle camera, where the core and the delta are usually located. In the second step, the correspondences which represent the same part of the skin should be provided. The CS calculated by the method introduced in Sect. 10.3.1 are then taken as the matched pairs in 2D fingerprint images. 3D space coordinates of such correspondences are also calculated by the method introduced in (Zhang 2000).

In the fourth step, to reconstruct the finger shape, we should get the shape model after certain 3D points of the finger are calculated. However, as an irregular 3D object, there is no exact model for human's finger shape which means the estimation of finger shape model is necessary. In our chapter, we estimate finger shape model by analyzing 440 3D point cloud finger data (220 fingers, 2 pictures each) collected by one camera and a projector using the SLI method. We found that the horizontal profile of such 3D finger shape is Parabola-like shape, while vertical profile of 3D finger shape can be represented by quadratic curve. Figure 10.15 shows an example of 3D point cloud of a finger. The green line represents the real data of horizontal profile and the red line is the fitting result by quadratic curve, while the blue line is the real line of vertical profile. Thus, the binary quadratic function $fxy = Ax^2 + By^2 + Cxy + Dx + Ey + F$ is taken as the finger shape model.

Finally, we reconstructed the 3D finger shape from fingerprint images captured by our device using the above mentioned steps. Figure 10.16 shows an example

Fig. 10.15 An example of 3D point cloud of a finger: *green line* depicts real data of horizontal profiles, *red line* is fitting by Parabola, *blue line* depicts real data of vertical profile

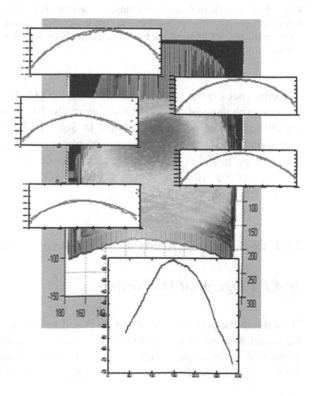

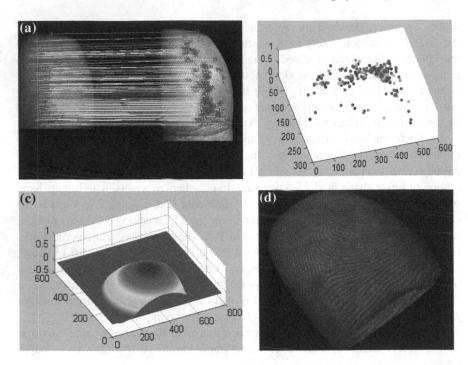

Fig. 10.16 Reconstruction of 3D finger shape of Fig. 10.11: **a** Correspondences established between Fig. 10.11a and Fig. 10.11d. **b** 3D space coordinates of matched pairs in (**a**). **c** Reconstructed 3D finger shape by fitting finger shape model. **d** 3D finger shape display with texture

to reconstruct the 3D finger shape from Fig. 10.11. Here, just the part above the distal interphalangeal crease is used. Figure 10.16a gives the correspondences we established by the method mentioned in Sect. 10.3.1. Figure 10.16b depicts the corresponding 3D space coordinates of matched pairs in Fig. 10.16a. After the finger shape model is used to fit such 3D space coordinates, we get the 3D finger shape of Fig. 10.11, as shown in Fig. 10.16c. Figure 10.16d shows the reconstructed finger shape with texture.

10.4 Performance Analysis and Comparison

10.4.1 Fingerprint Mosaicking

To evaluate the performance of the proposed mosaic method, we firstly compared the mosaicking result got by our method with the result obtained by the method mentioned in Choi's paper (Choi et al. 2010) in an example of images captured by the mirror-reflected device, as shown in Fig. 10.17. In the example image, there

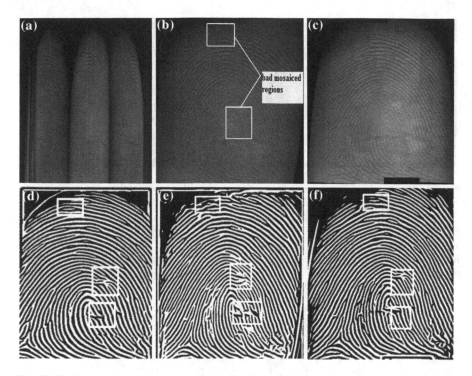

Fig. 10.17 Comparison of fingerprint mosaicking results: **a** Original image captured by the mirror-reflected device. **b** Mosaicked image obtained by using the method proposed by Choi et al. (2010). **c** Mosaicked image obtained by using the method proposed in this chapter. **d** Ridge map of original frontal image. **e** Ridge map of (**b**). **f** Ridge map of (**c**)

are bad mosaicked regions (labeled by red rectangles) using the method proposed in Choi's paper (Choi et al. 2010), while acceptable mosaicking result obtained by using the method proposed in this Chapter. Their corresponding ridge maps are also given in Fig. 10.17e and f. Compared with the ridge map of frontal image, the effective fingerprint areas are both enlarged. Better results in the edge and stitching regions are achieved using the proposed method. Our algorithm still works when images have very low ridge-valley contrast, Fig. 10.18 offers an example.

To verify that multi-view techniques can better solve the view difference problem caused by finger rolling and demonstrate that the mosaicked fingerprint image contains more useful information, we compared the recognition performance of fingerprint frontal images and mosaicked images in our established database. The database contains 541 fingers. Each finger has 4 pictures which were captured in two sessions (2 pictures per session).

The features and matching method introduced in Sect. 10.3.1 can also be used for fingerprint recognition. For simplicity, in this experiment, we extracted the SIFT feature and coarsely matched them by point wise matching method, and then refined them by RANSAC algorithm with TPS model. The size of CS is taken as

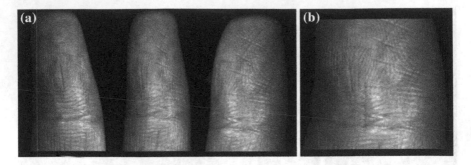

Fig. 10.18 Example of image mosaicking with very low ridge valley contrast: **a** Original image captured by our device. **b** Mosaicked image

Fig. 10.19 ROC curves for recognition with different fingerprint images

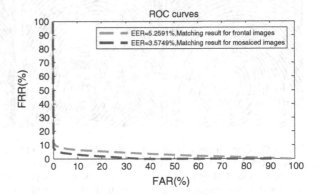

the match score. Figure 10.19 shows the ROC curves of recognition performance. Obviously, an EER of around 3.6 % is achieved by matching the mosaicked images, while an EER of around 5.3 % is obtained by matching the frontal images.

10.4.2 Comparison of Touch-Based Imaging and Touchless Multi-View Imaging

Generally, touch based fingerprint images has the advantage of high ridge-valley contrast but the disadvantage of small print size. While touchless multi-view fingerprint imaging technique permits large print size but low image quality. To find out whether multi-view technique compensates the drawbacks of touchless imaging at certain degree, we compared recognition accuracy of mosaicked images and touch based fingerprint images on small databases. Both of the touch based fingerprint images and the touchless multi-view fingerprint images are collected from same fingers. They both consist of 15 fingers, each two pictures. Figure 10.20 shows examples of our collected data. We still using the SIFT features and

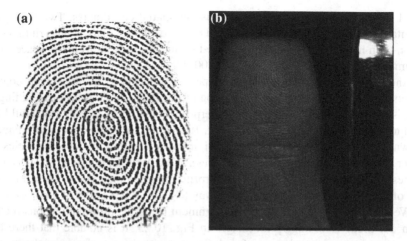

Fig. 10.20 Examples of fingerprint images from the same finger: **a** Touch based fingerprint image. **b** Frontal touchless fingerprint image

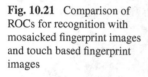

Fig. 10.21 Comparison of ROCs for recognition with mosaicked fingerprint images and touch based fingerprint images

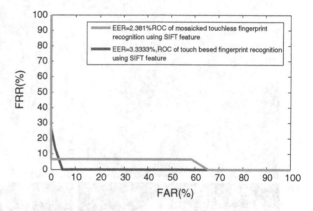

matching method introduced in Sect. 10.3.1. The size of CS is taken as the match score. Figure 10.21 shows the ROC curves of recognition performance. We found that there are comparable EERs by matching the mosaicked images and matching the touch based images.

10.4.3 3D Information Applied for Fingerprint Recognition

As mentioned in Parziale's paper (Parziale and Diaz-Santana 2006), one obvious advantage of adding 3D finger information is providing more attributes to 2D fingerprint features. For instance, a minutia feature in 2D fingerprint image is usually represented by its location $\{x, y\}$ and orientation θ. While in 3D case, it may be noted by $\{x, y, z, \theta, \phi\}$, where x, y and z is the spatial coordinates. Two angles of orientation of the ridge in 3D space θ and ϕ are available. Thus, fingerprint recognition with higher security can be achieved by matching features in 3D space (e.g. 3D minutia matching (Parziale and Niel 2004).

Another application of 3D finger shape we found is to do fingerprint coarse alignment which benefits fingerprint recognition. In 2D fingerprint images, finger skin is forced to plane. Features on fingerprint image (e.g. core point) should be used to align two images, as shown in Fig. 10.22a. While in 3D fingerprint images, shown in Fig. 10.22b, we can quickly align two images by the characteristics of finger shape, that the center part of the finger is higher than the side part. It is also interesting to find that core point in fingerprint images almost locate at the highest part of the finger, which also provided a way to locate the position of core point.

We gave an example of fingerprint alignment using 3D information constructed from our multi-view fingerprint images in Fig. 10.23. It is notable that there are two parts of reconstructed shape of each finger due to three cameras and the theory

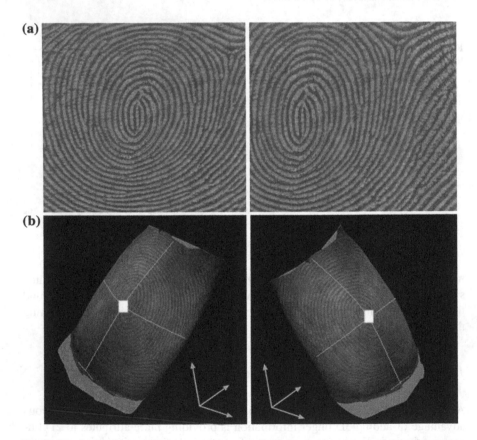

Fig. 10.22 Comparison of 2D fingerprint image and 3D fingerprint image in fingerprint coarse alignment: **a** Two 2D images of the same finger, **b** Two 3D images of the same finger

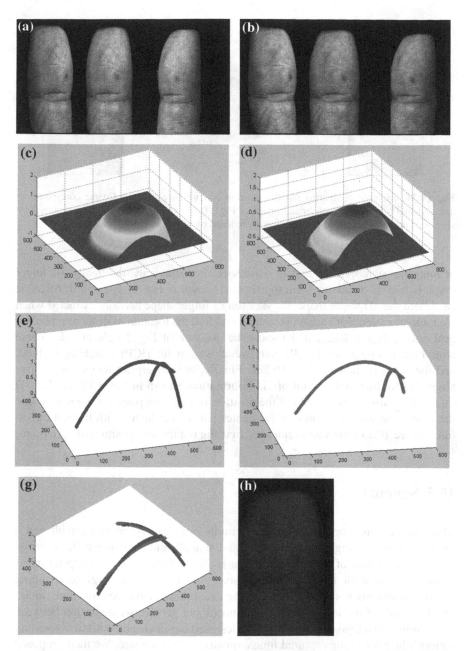

Fig. 10.23 3D Fingerprint coarse alignment: **a** Template of original touchless multi-view fingerprint images. **b** Test of original touchless multi-view fingerprint images. **c** Final reconstructed 3D finger shape of (**a**). **d** Final reconstructed 3D finger shape of (**b**). **e** 3D skeleton of (**c**). **f** 3D skeleton of (**d**). **g** 3D skeleton alignment by ICP. **h** Coarse alignment of template frontal-image and test frontal-image

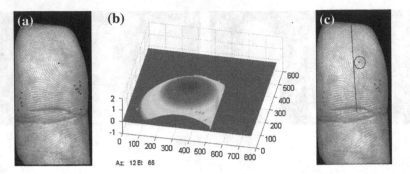

Fig. 10.24 Example of figuring out the position of true core point: **a** Coarse core point set detected by Poincare index on frontal image of Fig. 10.23a (*red point*). **b** 3D coarse core point set (*green point*). **c** Location of true core point indexed by 3D information (*red circled*)

of binocular stereo vision used in our device. For simplicity, the finally 3D finger shape is formed by adding them together, as shown in Fig. 10.23c and d. After we calculated the 3D finger shape, the skeleton of finger shape is easily extracted when we took the extrema of each cross section and find the location of distal interphalangeal crease. Figure 10.23e and f show the skeleton of Fig. 10.23c and d. Coarse alignment is completed by 3D iterative closest point (ICP) matching method, and the result is shown in Fig. 10.23g. Finally, two fingerprint images are coarse aligned according to the result of 3D information, shown in Fig. 10.23h. This 3D information also helps to find out the position of true core point for the reason that the core point usually locates on the center part of the finger with highest curvature. Figure 10.24 shows an example of figuring out the true position of core point.

10.5 Summary

This Chapter has proposed a touchless multi-view fingerprint and established a fingerprint mosaicking method, as well as the applications of finger 3D information. The advantage of multiview imaging is that it obtains more fingerprint information quickly while touchless imaging has the advantages of hygiene, avoiding fingerprint deformation, and not producing latent prints. However, touchless imaging does suffer from low ridge-valley contrast and perspective distortion between differently-posed images. Therefore, we designed our device by optimizing several factors which affect the captured image quality and device size. We then proposed a mosaicking method to get expanded fingerprint images with larger effective area. When mosaicking, we used the SIFT feature which is robust to low ridge-valley contrast. 3D finger shape feature reconstructed from multi-images is found to be useful to align two fingerprint images coarsely, which may benefit fingerprint matching. Experimental results show that the proposed mosaicking method well works on our captured images. An EER of 3.6 % by using SIFT feature in the mosaicked images can be achieved while an EER of 5.3 % is obtained in frontal images on our established database with 541 fingers. Comparable recognition accuracy is achieved when comparing EERs between mosaicked images and touch based images on smaller databases. Examples of fingerprint alignment by using the 3D information were for the first time shown in the Chapter.

Nonetheless, the current system still has some drawbacks which are inevitable to touchless imaging techniques. For some fingers, the image quality of the device is much lower than that of touch-based devices and some fingers are so tilted that some part of the fingerprint is out of the depth of field. There are either very narrow or wide ridges in one image due to the curve surface of finger. Such low quality fingerprints and large variations of ridge frequency call for a very robust feature extraction algorithm (e.g. minutiae extraction). Considering the fact that the area of touchless fingerprints is generally larger than that of touch-based fingerprints, we believe that future work will enable us to extract more distinctive information from touchless fingerprints than from touch-based fingerprints. We further foresee that the current system can be improved in the following three ways. First, a more robust feature extraction algorithm is required to deal with fingerprint images of very low quality and with large variations of ridge frequency. Second, we can obtain greater accuracy by the addition of non-minutiae information (e.g. finger shape, finger crease feature, image-based features, 3D information etc.). Finally, we propose to explore tighter fusion schemes, such as fusion at the feature or alignment level.

References

Bouguet JY (2010) Camera calibration toolbox for Matlab. http://www.vision.caltech.edu/bougu etj/calib_doc/index.html

Chen F (2006) 3D fingerprint and palm print data model and capture devices using multi structured lights and cameras. US Patent Application Publication, Pub. No. 2006/0120576

Chen Y (2009) Extended feature set and touchless imaging for fingerprint matching. PhD Dissertation, Michigan State University

Choi K, Choi H, Lee S, Kim J (2007) Fingerprint image mosaicking by recursive ridge mapping. Special issue on recent advances in biometrics systems. IEEE Trans Syst Man Cybern B 37(5):1191–1203. doi:10.1109/TSMCB.2007.907038

Choi H, Choi K, Kim J (2010) Mosaicing touchless and mirror-reflected fingerprint images. IEEE Trans Inf Forensics Secur 5(1):52–61. doi:10.1109/TIFS.2009.2038758

Elli A (2001) Understanding the color of human skin. In: Proceedings of 6th SPIE conference human vision and electronic imaging, SPIE 243–251. doi:10.1117/12.429495

Fatehpuria A, Lau D L, and Hassebrook L G (2006) Acquiring a 2D rolled equivalent fingerprint image from a non-contact 3D finger scan. In SPIE Defense and Security Symposium. Biometric Technology for Human Identification III, Orlando, FL pp 62020C-1–62020C-8. doi:10.1117/12.666127

Feng J (2008) Combining minutiae descriptors for fingerprint matching. Pattern Recogn 41(1):342–352. doi:10.1016/j.patcog.2007.04.016

Fishler M, Bolles R (1981) Random sample consensus: a paradigm for model fitting with applications to image analysis and automated cartography. Commun ACM 24(6):381–395. doi:10.1145/358669.358692

Hartley R, Zisserman A (2003) Multiple view geometry in computer vision. Cambridge University Press, UK. ISBN 0521540518

Jain A, Ross A (2002) Fingerprint mosaicking. In: Proceedings of IEEE international conference on acoustics, speech, and signal processing (ICASSP) pp 13–17. doi:10.1109/ICASSP.2002.5745550

Kumar A, Zhou YB (2011) Contactless fingerprint identification using level zero features. In: Proceedings CVPR'11, CVPR'W pp 121–126. doi:10.1109/CVPRW.2011.5981823

Lowe DG (2004) Distinctive image features from scale-invariant keypoints. Int J Comput Vision 60(2):91–110. doi:10.1023/B:VISI.0000029664.99615.94

Lumidigm multispectral fingerprint imaging (2012) http://www.lumidigm.com

Malathi S, Meena C (2010) Partial fingerprint matching based on SIFT features. Int J Comput Sci Eng 2(4):1411–1414

Mitsubishi touchless fingerprint sensor (2012) http://global.mitsubishielectric.com

Parziale G, Diaz-Santana E (2006) The surround imager: a multi-camera touchless device to acquire 3D rolled-equivalent fingerprints. In: Proceedings of international conference on biometrics (ICB), Hong Kong, China pp 244–250. doi:10.1109/BCC.2006.4341621

Parziale G, Niel A (2004) A fingerprint matching using minutiae triangulation. In: Proceedings of international conference on biometric authentication (ICBA), LNCS pp 241–248. doi:10.1007/978-3-540-25948-0_34

Pankanti S, Prabhakar S, Jain AK (2002) On the individuality of fingerprints. IEEE Trans Pattern Anal Mach Intell 24(8):1010–1025. doi:10.1109/TPAMI.2002.1023799

Park U, Pankanti S, Jain AK (2008) Fingerprint verification using SIFT features. In: Proceedings of SPIE6944, pp 69440K–69440K-9. doi:10.1117/12.778804

Ross A, Dass S, Jain AK (2005) A deformable model for fingerprint matching. Pattern Recogn 38(1):95–103. doi:10.1016/j.patcog.2003.12.021

Shah S, Ross A, Shah J, and Crihalmeanu S (2005) Fingerprint mosaicing using thin plate splines. In: The biometric consortium conference

Song Y, Lee C, Kim J (2004) A new scheme for touchless fingerprint recognition system. In: Proceedings of the 2004 international symposium on intelligent signal processing and communication systems, pp 524–527. doi:10.1109/ISPACS.2004.1439111

Sonka M, Hlavac V, Boyle R (1999) Image processing, analysis, and machine vision, 2nd edn. Brooks/Cole Publishing, USA. ISBN 053495393X

TBS Touchless Fingerprint Imaging: 3D-enroll, 3D-terminal (2012) http://www.tbsinc.com/

TST Biometrics BiRD 3 (2012) http://www.tst-biometrics.com

Wang L, H R, El-Maksoud Abd, Sasian JM and Valencia VS (2009) Illumination scheme for high-contrast, contactless fingerprint images. In: Proceedings of SPIE, vol 7429, p 742911. doi:10.1117/12.828523

Xia X, O'Gorman L (2003) Innovations in fingerprint capture devices. Pattern Recogn 36(2):361–369. doi:10.1016/S0031-3203(02)00036-5

Zhang Z (2000) A flexible new technique for camera calibration. IEEE Trans Pattern Anal Mach Intell 22(11):1330–1334. doi:10.1109/34.888718

Zhang D, Liu F, Zhao Q, Lu G, Luo N (2011) Selecting a reference high resolution for fingerprint recognition using minutiae and pores. IEEE Trans Instrum Meas 60(3):863–871. doi:10.1109/TIM.2010.2062610

Chapter 11
3D Fingerprint Reconstruction

Abstract As a 3D object, human finger provides more real and more features for personal identification if 3D fingerprint images are available. This chapter thus studies 3D fingerprint reconstruction technique from touchless multi-view fingerprint images captured by our own designed device, which offers a solution for 3D fingerprint recognition. The reconstruction technique follows binocular stereo vision theory in computer vision domain, which mainly contains camera parameters calculation, correspondences establishment, 3D coordinates computation, and shape model estimation. For 3D fingerprint reconstruction, the difficulties and stresses focus on correspondence establishment from 2D touchless fingerprint images and finger shape model estimation. In this chapter, several popular used features, such as scale invariant feature transformation (SIFT) feature, ridge feature and minutia, are considered for correspondence establishment. To extract such fingerprint features accurately, we propose improved fingerprint enhancement method by polishing orientation map and ridge frequency map specific to the characteristics of 2D touchless fingerprint images. Correspondences are finally established by adopting hierarchical fingerprint matching approaches. Finger shape model in our chapter is estimated by analyzing 440 3D point cloud finger data collected by one camera and a projector using the structured light illumination (SLI) method. We found binary quadratic function is more suitable for finger shape model compared with another mixed model we used in this chapter. Experimental results firstly show the reconstruction accuracy of our device by constructing a cylinder. 3D fingerprint reconstruction results from different fingerprint feature correspondences are then given and the reconstruction accuracy is finally analyzed and compared.

Keywords 3D fingerprint reconstruction • Touchless multi-view: imaging • SIFT • Minutia • Ridge map • 3D finger shape model

D. Zhang and G. Lu, *3D Biometrics*, DOI: 10.1007/978-1-4614-7400-5_11,

11.1 Introduction

The 3D information of an object can be obtained from its two different plane pictures captured at one time according to binocular stereo vision theory in computer vision domain. As shown in Fig. 11.1a, given two images Cl and Cr captured at one time, the 3D coordinate of A can be calculated if some camera parameters (f_l, f_r, O_l, O_r etc.) and matched pair (($a_l (u_l, v_l)$) ↔ ($a_r (u_r, v_r)$)) are provided. Once we know the shape model and several calculated 3D coordinates of the 3D object, the shape of the 3D object can be obtained after interpolation. Like Fig. 11.1b shown, the triangle in 3D space is obtained after computing 3D coordinates of three vertices and shape fitting by triangle model.

Fortunately, our touchless multi-view fingerprint imaging device had been designed to provide three images of the finger simultaneously, as shown in Fig. 10.9, which means 3D finger shape is available based on stereovision and photogrammetry methods. As stated in the previous paragraph, there are mainly four steps to reconstruct a 3D object. That is camera parameters calculation, correspondences establishment, 3D coordinates computation, and shape model estimation.

The differences between other 3D objects and 3D fingerprint reconstruction are focus on correspondences establishment and shape model estimation. There are a few works for touchless fingerprint matching due to the characteristics of touchless fingerprint imaging and hardly any work can be found for finger shape model analysis. Thus, in this chapter, we for the first time studied model of human finger and analyzed touchless fingerprint features for correspondences establishment, and then reconstructed 3D finger shape based on the images provided by our device.

11.2 3D Fingerprint Reconstruction from Multi-View Images

Among four steps of 3D reconstruction, the emphasis is on matched pairs founding in 2D pictures and 3D shape model definition for different kinds of 3D objects. We then introduced the procedures of 3D fingerprint reconstruction step by step

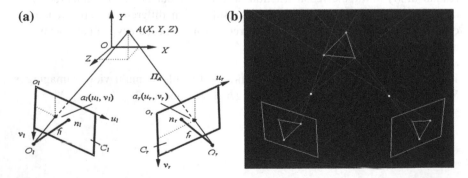

Fig. 11.1 An illustration of constructing a 3D triangle by binocular stereo vision: **a** 3D coordinates calculation on 3D space, **b** 3D triangle reconstruction

and focused on correspondences establishment between two fingerprint images and finger shape model estimation method.

11.2.1 Camera Calibration

Camera calibration is the first step for 3D reconstruction. It provides the intrinsic parameters (Focal Length, Principal Point, Skew, and Distortion) of each camera and extrinsic parameters (Rotation, Translation) between cameras necessary for reconstruction. It usually implemented off-line. In this chapter, the methodology proposed in Zhang's paper (Zhang 2000) and the improved algorithm coded by Bouguet (2010) is employed. It is notable that there are three cameras used in our fingerprint captured device. The position of the middle camera is chosen as the reference system for the reason that central part of the fingerprint is more likely to be captured by the middle camera, where the core and the delta are usually located. We also asked volunteers to collaborate when collecting.

11.2.2 Correspondence Establishment for Touchless Fingerprints

Fingerprints are distinguished with their features. Different fingerprint features can be observed for different resolution fingerprint images. Currently, there are mainly three level of fingerprint features popular used in fingerprint literatures (Jain et al. 2007; Zhang et al. 2011), as shown in Fig. 11.2. Level 1 features are defined as the macro details of fingerprints such as singular points and global ridge patterns (deltas and cores indicated by red triangles in Fig. 11.2). The level 2 features (red rectangles) primarily refer to the Galton features or minutiae, namely ridge endings and bifurcations. Level 3 features (red circles) are the dimensional attributes of the ridges including sweat pores, ridge contours, and ridge edge features.

However, for touchless fingerprint images usually captured with low resolution, high level fingerprint features are difficult to extract. Kumar then added level 0

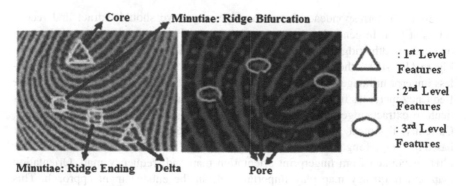

Fig. 11.2 Three levels of fingerprint features

features that are Localized texture patterns for low resolution touchless fingerprint recognition (Kumar and Zhou 2011). Low level fingerprint features seem advisable for touchless fingerprint correspondence establishment for the reason that high level fingerprint features are very difficult to extract from touchless fingerprint images. Meanwhile, we do not need to recognize personal identification but just need matched pairs between different views of same finger for 3D reconstruction. In general, there are three frequently-used features, namely SIFT feature, ridge map and minutiae adopted in fingerprint matching and mosaicing domain (Choi et al. 2010; Park et al. 2008; Feng 2008; Malathi and Meena 2010; Jain and Ross 2002; Shah et al. 2005; Choi et al. 2007). This chapter thus tries to extract such features and establish correspondences between different views of fingerprint images.

11.2.2.1 Correspondence Establishment Based on SIFT Feature

Scale Invariant Feature Transformation (SIFT) (Lowe 2004) was quite popular in object recognition and image retrieval. It provides feature which is invariant to scale, rotation and affine transforms. We adopted this feature in our algorithm for the following reasons. Firstly, it is inevitable to capture fingerprint images with very low ridge-valley contrast by touchless imaging techniques. It permits true correspondences can be established when minutiae and ridge features cannot correctly extract. Secondly, SIFT feature depicts local texture pattern in pixel level which means pixel accuracy can be reached for correspondences establishment. Thirdly, SIFT feature is robust to deformation variation and rich in quantity (Park et al. 2008; Malathi and Meena 2010). Figure 11.3b and d shows examples of our extracted SIFT features. Their total number is as much as 1,911 and 1,524, respectively. The method to extract match SIFT feature are already introduced in last Sect. 10.3.1.

11.2.2.2 Correspondence Establishment Based on Ridge Map

To establish correspondence between ridge maps, we should extract and record ridges at first. In general, ridge map refers to the thinning image where ridges are one-pixel-width, ridge pixels have value 1 and background pixels have value 0. Figure 11.4 shows the flowchart of steps for ridge map extraction. However, touchless fingerprint images has the characteristics of low ridge-valley contrast and ridge frequency increases from center to side, as Fig. 11.3a and c shown. These make difficult to extract ridge map accurately due to the hard of fingerprint enhancement. Currently, the most famous fingerprint enhancement method is the Gabor filter based method (Hong et al. 1998). Fingerprint image is enhanced by a bank of Gabor filters generated from fingerprint orientation map and frequency map. Orientation map and frequency map play important role in the enhancement approach. This

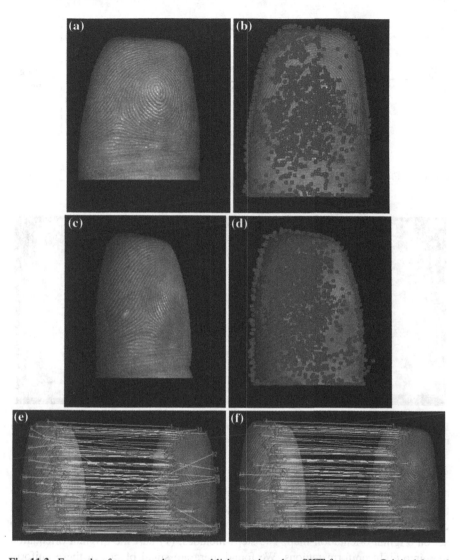

Fig. 11.3 Example of correspondences establishment based on SIFT features: **a** Original frontal image, **b** Extracted SIFT feature from (**a**), **c** Original *left-side* image, **d** Extracted SIFT feature from (**c**), **e** Initial correspondences establishment by point wise matching, **f** Final correspondences after refining by RANSAC method

chapter thus tried to improve the orientation map and frequency map so as to get better enhanced results.

As is introduced in Maltoni's paper (Maltoni et al. 2009), gradient-based ridge orientation estimation method is the simplest and most intuitive one. It is efficient and popularly used in fingerprint recognition domain. However, it also has drawbacks, such as sensitive to noise when orientation estimated at too fine a scale or

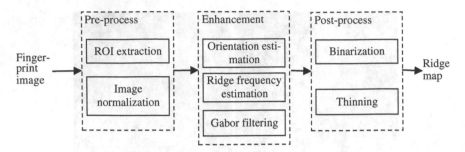

Fig. 11.4 Flowchart of ridge map extraction

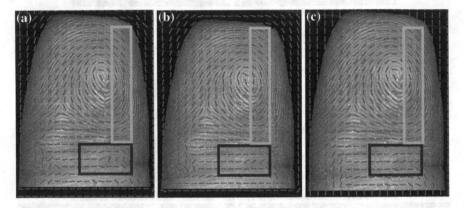

Fig. 11.5 Fingerprint ridge orientation maps. **a** Original orientation map. **b** Smoothed orientation map of (**a**). **c** Orientation map post-processed by our proposed method

accuracy decreased if smooth factor used, as Fig. 11.5a (red rectangle) and Fig. 11.5b (green rectangle) shown. To keep the estimation accuracy in good quality region and correct the orientation where noises exist, we proposed a post-processing method to act on original orientation map. The steps includes: (1) Part the original orientation map into eight uniform regions. We found there are small blocks in these uniform regions where represents wrong orientation estimation result (see Fig. 11.6a, red circled); (2) Sort uniform regions with the same color in a descending manner, such regions whose size is smaller than the mean size of all regions with the same color are set to zero (see Fig. 11.6b, dark regions in ROI); (3) Assign values to the points with zero value set by step (2) according to nearest neighbor method. The improved orientation map is obtained by following these three steps. Figure 11.5c shows our improved orientation map based on Figs. 11.5a and 11.6c gives the partition map according to Fig. 11.5c. We can see that it keeps the estimation accuracy in good quality region and correct the orientation where noises exist (Fig. 11.5c, rectangle).

Frequency map records the number of ridges per unit length along a hypothetical segment and orthogonal to the local ridge orientation. The simplest and most

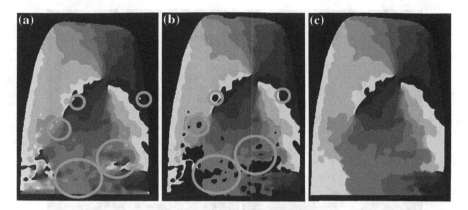

Fig. 11.6 Partition results according to orientation map. **a** Partition result according to original orientation map. **b** Partition result according to our improved orientation map map. **c** Partition map

popular ridge frequency estimation method is x-signatures based method (Maltoni et al. 2009). However, this kind of method did not work in blurry or noisy fingerprint regions. In this situation, interpolation and filtering is used to post-process the original estimated frequency map. For touchless fingerprint images, frequency map is harder to be estimated than touch based fingerprint images due to the low ridge-valley contrast of touchless fingerprint images and simple interpolation or filtering is invalid when wrong frequency estimated in neighborhoods. Fortunately, we found that frequency increases or period decreases from central part to side part for horizontal section and frequency decreases from finger tip to distal interphalangeal crease for vertical section for touchless fingerprint image, as Fig. 11.7 shown (ridge frequency is calculated with blocks of 32 * 32 pixels). This phenomenon can be explained from touchless capturing technique and the observation of human finger. As Eq. (11.1), M is the optical magnification. p and q are the lens-to-object and lens-to-image distances, respectively. The larger the magnification M is, the larger the ridge period (1/ridge frequency) is. For a fix p, large q corresponds to small magnification M. Figure 10.7 illustrates the distribution of q (D1, D2, and D3). Thus, it is larger in the central part of ridge period than side-view ones for horizontal section. For vertical distribution of ridge period, it is increasing from finger tip to distal interphalangeal crease because q increases from tip to center part of the finger and ridges are wider near distal interphalangeal crease than other parts by observation,

$$M = \frac{q}{p}. \tag{11.1}$$

According to the distribution of ridge frequency of touchless fingerprint images, we proposed to use monotone increasing function (logarithmic function) to fit ridge period (1/ridge frequency) map along vertical direction and quadratic curve along horizontal direction. The ridge period map is finally fitted by a mixed model of logarithmic function and quadratic curve. Equation (11.2) shows the

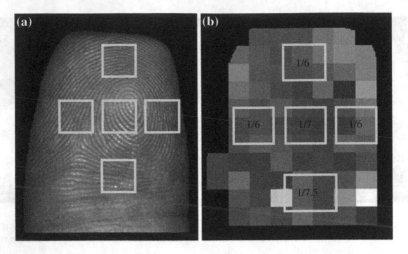

Fig. 11.7 Frequency variation of touchless fingerprint images. **a** Original touchless fingerprint image. **b** Corresponding frequency map

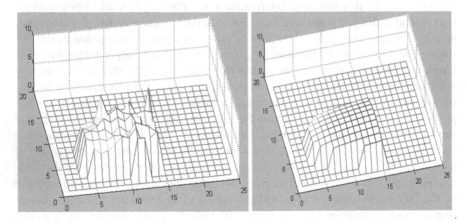

Fig. 11.8 Ridge period distribution maps of Fig. 11.8. Original ridge period map (*left*), our improved ridge period map (*right*)

formulation of the mixed model. Figure 11.8 shows an example of original ridge period map and our improved one. We can see that our improved ridge period map is closer to our observation of ridge distribution than original one.

$$fxy = Ax^2 + Bx + C\ln(y) + D \tag{11.2}$$

Once we got the orientation and ridge frequency map, a series of Gabor filter was generated based on them. The enhanced fingerprint image was then obtained, as shown in Fig. 11.9c. After binarizing the enhanced fingerprint image by simple threshold method and morphology approach, we got the final ridge map, as shown in Fig. 11.9d. Figure 11.9a and b shows the enhanced fingerprint image using

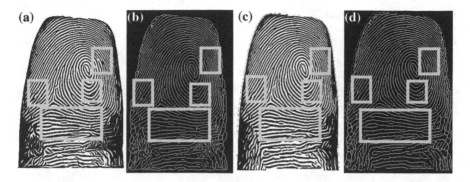

Fig. 11.9 Ridge maps. **a** Enhanced image by original orientation and ridge frequency map. **b** Thinned ridge map of (**a**). **c** Enhanced image by improved orientation and ridge frequency map. **d** Thinned ridge map of (**c**)

original orientation map and original ridge frequency map interpolated by mean value of frequency map. We found we got better results using our improved orientation and ridge frequency map when compared Fig. 11.9c, d with Fig. 11.9a, b (red rectangles). It is notable the pre-process steps of ROI extraction and normalization followed the method proposed in references (Hong et al. 1998; Maltoni et al. 2009).

Before correspondences establishment, ridges are record by tracing starting from minutiae where ridges are disconnected. Due to existence of noise, ridge image often has some spurs and breaks. In some cases, the noise is not strong and correct ridge structure can be recovered by removing short ridges or connecting broken ridges. However, in other cases, the noise is too strong to recover the correct ridge structure by removing short ridges or connecting broken ridges. In such cases, we remove all related ridges. Finally, the down sampled ridge point coordinates of each ridge is recorded in a list.

Coarse alignment of two ridge maps is done by using the global transform model using the method introduced in Sect. 10.3.1 when SIFT feature matched. Ridges in ridge maps are then matched by adopting Dynamic Programming (DP) method. As shown in Fig. 11.10 and Table 11.1, for any ridge in template and test ridge maps, calculated the Euclidian distance between each pair of compared ridge lines. The status will be 1 if the distance of a pair of ridge points is smaller than a threshold (we set 5 points in this chapter), otherwise, the status will be 0. The DP method is adopted to find matched ridge pairs with largest number. Coarse ridge correspondences are then established after DP, we also used RANSAC algorithm introduced in Sect. 10.3.1 to select true ones from the coarse set. Figure 11.11 shows the results we got of ridge correspondence establishment.

11.2.2.3 Correspondence Establishment Based on Minutiae

As we know, minutiae are widely used for fingerprint recognition due to their strong discriminative ability. We found the distribution of minutiae is covered most of fingerprint images. This feature thus considered in our chapter.

Fig. 11.10 Correspondences establishment between two ridges

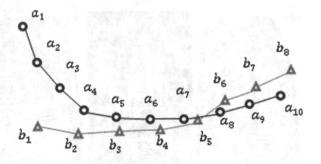

Table 11.1 Record of status among ridge points in Fig. 11.10

	a_1	a_2	a_3	a_4	a_5	a_6	a_7	a_8	a_9	a_{10}
b_1	0	0	0	0	0	0	0	0	0	0
b_2	0	0	0	0	0	0	0	0	0	0
b_3	0	0	0	0	1	0	0	0	0	0
b_4	0	0	0	0	0	1	1	0	0	0
b_5	0	0	0	0	0	0	1	1	0	0
b_6	0	0	0	0	0	0	0	1	0	0
b_7	0	0	0	0	0	0	0	0	1	0
b_8	0	0	0	0	0	0	0	0	0	0

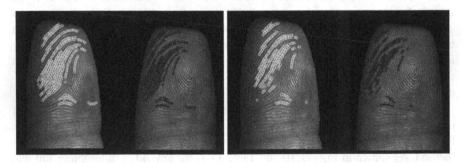

Fig. 11.11 Ridge correspondence establishment: Initial correspondences (*left*), Final correspondences after RANSAC (*right*)

Minutiae can be easily extracted in our chapter since ridge map is obtained. Figure 11.12 shows an example of extracted minutiae based on the ridge map we got by using the method mentioned in Sect. 11.2.2 and the minutiae extraction method introduced in Jain's paper (Jain et al. 1997).

Since we got the transformation model when establishing SIFT correspondence, we can easily establish initial minutiae correspondence by nearest neighbour method and then get final results by RANSAC with TPS model, as shown in Fig. 11.13.

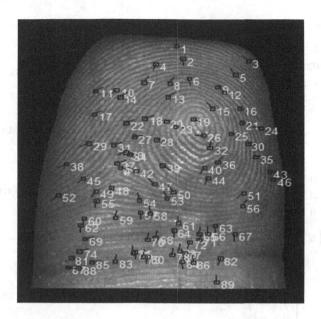

Fig. 11.12 Extracted minutiae result

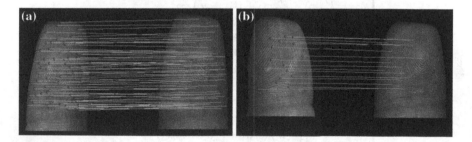

Fig. 11.13 Minutiae correspondences establishment. **a** Initial correspondences. **b** Final correspondences after RANSAC

11.2.2.4 3D Coordinate Calculation for Matched Fingerprint Points

Once camera parameters and matched pairs between different view fingerprint images are both obtained, the 3D coordinate of each correspondence can be calculated by solving Eq. (11.3),

$$x^* = \left(A^T A\right)^{-1} A^T B, \tag{11.3}$$

where $x^* = [X\,Y\,Z]^T$ is the 3D coordinate needs to be calculated;

$A = \begin{bmatrix} a_{11} & a_{12} & a_{13} \\ a_{21} & a_{22} & a_{23} \\ a_{31} & a_{32} & a_{33} \\ a_{41} & a_{42} & a_{43} \end{bmatrix}$ and $B = [b_1\,b_2\,b_3\,b_4]^T$ are two matrix consisting of intrinsic

parameters $K = \begin{bmatrix} k_{11} & k_{12} & k_{13} \\ k_{21} & k_{22} & k_{23} \\ k_{31} & k_{32} & k_{33} \end{bmatrix}$, $K' = \begin{bmatrix} k'_{11} & k'_{12} & k'_{13} \\ k'_{21} & k'_{22} & k'_{23} \\ k'_{31} & k'_{32} & k'_{33} \end{bmatrix}$, rotation matrix

$R_{3\times3} = \begin{bmatrix} r_{11} & r_{12} & r_{13} \\ r_{21} & r_{22} & r_{23} \\ r_{31} & r_{32} & r_{33} \end{bmatrix}$, translation vector $T_{3\times1} = [t_1\,t_2\,t_3]^T$ and correspondence

$([u\;v] \leftrightarrow [u'\;v'])$ in two images. More specifically,

$$
\begin{cases}
a_{11} = k_{11} \\
a_{12} = k_{12} \\
a_{13} = k_{13} - u \\
a_{21} = k_{21} \\
a_{22} = k_{22} \\
a_{23} = k_{23} - v \\
a_{31} = k'_{11}r_{11} + k'_{12}r_{12} + (k'_{13} - u')\,r_{13} \\
a_{32} = k'_{11}r_{21} + k'_{12}r_{22} + (k'_{13} - u')\,r_{23} \\
a_{33} = k'_{11}r_{31} + k'_{12}r_{32} + (k'_{13} - u')\,r_{33} \\
a_{41} = k'_{21}r_{11} + k'_{22}r_{12} + (k'_{23} - v')\,r_{13} \\
a_{42} = k'_{21}r_{21} + k'_{22}r_{22} + (k'_{23} - v')\,r_{23} \\
a_{43} = k'_{21}r_{31} + k'_{22}r_{32} + (k'_{23} - v')\,r_{33} \\
b_1 = 0 \\
b_2 = 0 \\
b_3 = k'_{11}\,(r_{11}t_1 + r_{21}t_2 + r_{31}t_3) + k'_{12}\,(r_{12}t_1 + r_{22}t_2 + r_{32}t_3) \\
\qquad + (k'_{13} - u')\,(r_{13}t_1 + r_{23}t_2 + r_{33}t_3) \\
b_4 = k'_{21}\,(r_{11}t_1 + r_{21}t_2 + r_{31}t_3) + k'_{22}\,(r_{12}t_1 + r_{22}t_2 + r_{32}t_3) \\
\qquad + (k'_{23} - v')\,(r_{13}t_1 + r_{23}t_2 + r_{33}t_3)
\end{cases}
$$

11.2.3 Finger Shape Model Estimation

To reconstruct the finger shape, we should get the shape model after certain 3D points of the finger are calculated. However, as an irregular 3D object, there is no exact model for human's finger shape which means the estimation of finger shape model is necessary. In our chapter, we estimate finger shape model by analyzing 440 3D point cloud finger data (220 fingers, 2 pictures each) collected by one camera and a projector using the SLI method. Figure 11.14a–c display the 3D point cloud data of a thumb, index and little fingers we collected. We then randomly selected and drew the horizontal profile and vertical profile of these 3D point cloud data. We found that the horizontal profile of such 3D finger shape is Parabola-like shape, as

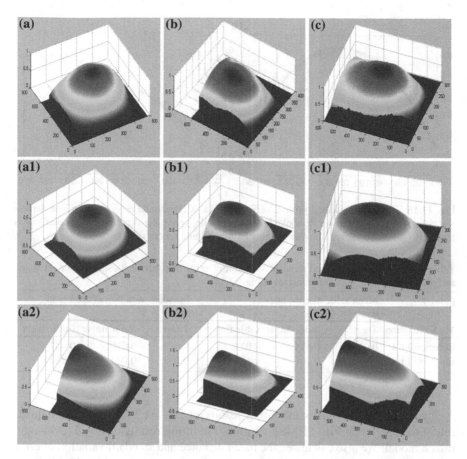

Fig. 11.14 Representative 3D finger shape data and their fitting results by different models. **a** 3D point cloud data of a thumb. **b** 3D point cloud data of a index finger. **c** 3D: point cloud data of a little finger. **a1** Fitting results of (**a**) by binary quadratic function. **b1** Fitting results of (**b**) by binary quadratic function. **c1** Fitting results of (**c**) by binary quadratic function. **a2** Fitting results of (**a**) by a mixed model with parabola and logarithmic function. **b2** Fitting results of (**b**) by a mixed model with parabola and logarithmic function. **c2** Fitting results of (**c**) by a mixed model with parabola and logarithmic function

shown in Fig. 11.15a, the green line depicts the real data and the red line is the fitting result by parabola. While vertical profile of 3D finger shape can be represented by a quadratic curve or a logarithmic function (see Fig. 11.15b), the green line represents the real data, red line is the fitting result by quadratic curve, and green line is the fitting result by logarithmic function. Thus, the binary quadratic function

$$f(x, y) = Ax^2 + By^2 + Cxy + Dx + Ey + F, \qquad (11.4)$$

and mixed model with parabola and logarithmic function

$$f(x, y) = Ax^2 + Bx + C\ln(y) + D, \qquad (11.5)$$

are used to fitting 440 3D point cloud finger data by regression method.

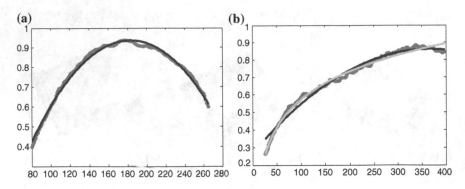

Fig. 11.15 Randomly selected profiles of Fig. 11.14a. **a** Horizontal profile, green line depicts real data, *red line* is fitting by Parabola. **b** Vertical profile, *green line* depicts real data, *red line* is fitting by *Quadratic Curve*, *blue line* is fitting by logarithmic function

Table 11.2 Mean distance and standard variation of error map between estimated finger shape and real finger shape of example images in Fig. 11.14

Index fitting	Mean value of error map $mean\,(\tilde{V} - V)$			Standard variation of error map $std\,(\tilde{V} - V)$		
Model function	Thumb	Index finger	Little finger	Thumb	Index finger	Little finger
Equation 11.4	0.024	0.036	0.037	0.019	0.029	0.027
Equation 11.5	0.082	0.061	0.103	0.057	0.042	0.062

Figure 11.14a1–c1 gives the fitting results of Fig. 11.14a–c by binary quadratic function, while Fig. 11.14a2–c2 gives the fitting results of Fig. 11.14a–c by mixed model. We also calculated the mean distance and standard variation of error map between estimated finger shape and real finger shape of Fig. 11.14. From Table 11.2, we can see that the binary quadratic function is closer to real finger shape than the mixed model. Then, the mean distance and standard variation of error map between estimated finger shape and real finger shape of all 440 fingers we collected are computed. Figure 11.16 shows the distributions of them. It can be seen that the binary quadratic function is more suitable for finger shape model. We finally chose to use binary quadratic function as the finger shape model.

11.3 Experimental Results and Analysis

11.3.1 Reconstruction Technique and Capture System Error Analysis

It is inevitable there are reconstruction and system errors. To test these errors, we firstly try to reconstruct an object with standard cylinder shape and of radius 10 mm.

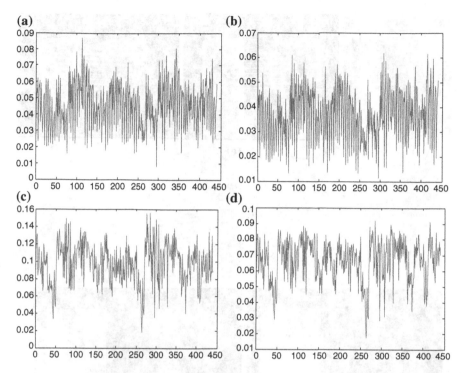

Fig. 11.16 Distributions of the mean distance and standard variation of error map between estimated finger shape and real finger shape of all 440 fingers we collected. **a** Distribution of the mean distance of error map between fitting result by binary quadratic function and real finger shape of all 440 fingers. **b** Distribution of the standard variation of error map between fitting result by binary quadratic function and real finger shape of all 440 finger. **c** Distribution of the mean distance of error map between fitting result by mixed model and real finger shape of all 440 fingers. **d** Distribution of the standard variation of error map between fitting result by mixed model and real finger shape of all 440 fingers

The surface of the object is wrapped by a grid paper to facilitate feature extraction, as shown in Fig. 11.17a. Figure 11.17b describes correspondences between left-side and frontal images and Fig. 11.17c shows correspondences between right-side and frontal images. It is notable that the corner features of grid are selected and correspondences between images are established manually. The result of 3D coordinates of such correspondences (norm to [0, 1] by min–max strategy) is shown in Fig. 11.17d and e respectively. From Fig. 11.17f and Fig. 11.18g, we can see that the radius of reconstructed cylinders from 40 3D points of Fig. 11.17d and e are −9.91 and −9.84 mm compared with the real radius 10 mm. Figure 11.17h and i give the error maps of 3D points corresponding to Fig. 11.17d and e when fitting by cylinder shape with radius of

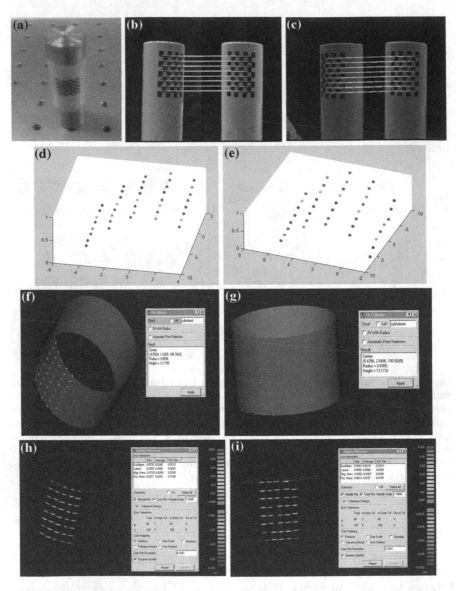

Fig. 11.17 Reconstruction accuracy analysis of cylinder shape object. **a** Original cylinder shape object wrapped with grid paper. **b** Correspondences established between *left-side* and frontal images captured by our device. **c** Correspondences established between *right-side* and frontal images captured by our device. **d** 3D space points corresponding to (**b**). **e** 3D space points corresponding to (**c**). **f** Fitting result corresponding to (**d**). **g** Fitting result corresponding to (**e**). **h** Error map corresponding to (**d**) when fitting by cylinder shape with radius of 10 mm. **i** Error map corresponding to (**e**) when fitting by cylinder shape with radius of 10 mm

10 mm. The error ranges are (−0.07 to 0.06 mm) and (−0.1 to 0.06 mm) separately. These results demonstrate that the reconstruction error of our device is within −0.2 mm.

11.3.2 3D Fingerprint Reconstruction Results and Analysis

By following the four steps we introduced in Sect. 11.2, reconstructed 3D finger-print images are obtained. Since there are three fingerprint images captured at one time and the central camera is selected as the reference system, our system consists of two parts (left-side camera and central camera, right-side camera and central camera) according to binocular stereo vision theory. Thus, we combined two parts of our system before the fourth steps by normalizing the calculated depth value of correspondences into [0 1]. Here, the Min–Max strategy of normalization is used. This combination is adopted for two reasons. One is there are parts of overlapping region between two adjacent fingerprint images, the distribution of correspondences may focus on a small part of fingerprint images. Larger area of fingerprint image can be covered by discrete correspondences through combining two parts of our system. The other one is it is very simple to accomplish and system error of combining two parts before model fitting is alleviated.

Table 11.3 then shows the reconstruction results based on three different finger-print feature correspondences we used of an example images shown in Fig. 11.18. We can see that the results are different corresponding to different feature matched pairs due to quite different numbers and distribution of established fingerprint feature correspondences and the existence of false correspondences.

To investigate which features are more suitable for 3D fingerprint reconstruction, we also manually labeled fingerprint correspondences, as shown in Fig. 11.19. We plotted the histogram of error map between reconstructed results in Table 11.3 and Fig. 11.19 in Fig. 11.20. From Fig. 11.20, we can see that for single feature used, reconstruction result based on SIFT features achieves best result while ridge feature based is the worst one. When combining other features, the combination of SIFT feature and minutiae can get best result for 3D fingerprint reconstruction. We also compared the reconstruction results with the 3D point cloud collected by one camera and a projector using the SLI method, as shown

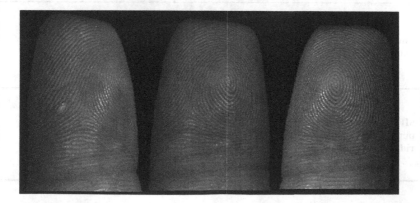

Fig. 11.18 Example fingerprint images captured by our device (*left, middle, right*)

Table 11.3 Reconstruction results from different feature correspondences of Fig. 11.18

Results Used feature	Established correspondences	Reconstructed 3D fingerprint image
SIFT feature		
Minutiae		
Ridge feature		
Feature Combination	Reconstructed 3D fingerprint image	
SIFT feature and minutiae		
SIFT and ridge feature		
Minutiae and ridge feature		
SIFT feature, minutiae and ridge feature		

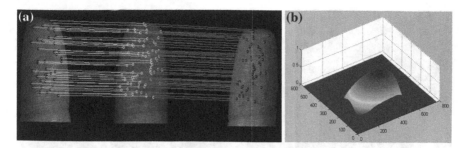

Fig. 11.19 Reconstruction of 3D finger shape of Fig. 11.18. **a** Manually labeled correspondences between fingerprint images. **b** Reconstructed 3D finger shape based on (**a**)

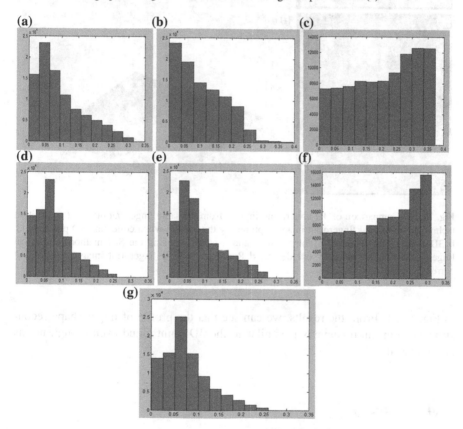

Fig. 11.20 Histogram of error map between reconstructed results in Table 11.3 and Fig. 11.19. **a** Histogram of err map between Fig. 11.19b and reconstruction result by using SIFT feature only. **b** Histogram of err map between Fig. 11.19b and reconstruction result by using minutiae only. **c** Histogram of err map between Fig. 11.19b and reconstruction result by using ridge feature only. **d** Histogram of err map between Fig. 11.19b and reconstruction result by using both SIFT feature and minutiae. **e** Histogram of err map between Fig. 11.19b and reconstruction result by using both SIFT feature and ridge feature. **f** Histogram of err map between Fig. 11.19b and reconstruction result by using both minutiae and ridge feature. **g** Histogram of err map between Fig. 11.19b and reconstruction result by using SIFT feature, minutiae and ridge feature

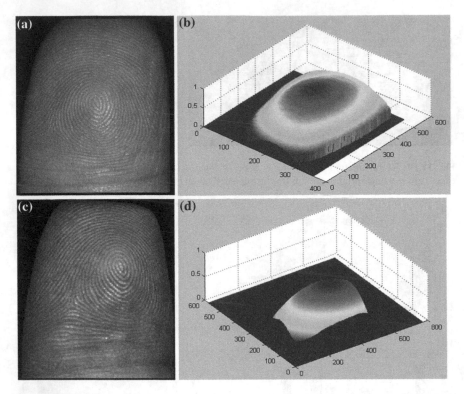

Fig. 11.21 Comparison of 3D fingerprint: images from the same finger but different acquisition technique. **a** Original fingerprint image captured by the camera when collecting 3D point cloud. **b** 3D point cloud collected by one camera and a projector using the SLI method. **c** Original fingerprint image captured by our device. **d** Reconstructed 3D fingerprint image with labeled correspondences

in Fig. 11.21. From the results, we can see that the profile of finger shape reconstructed from multi-cameras is similar to the 3D point cloud even though not as accurate as it.

11.4 Summary

This chapter has proposed a method for 3D fingerprint reconstruction based on touchless multi-view fingerprint images, which provides a way to achieve 3D fingerprint recognition. For 3D fingerprint reconstruction, the difficulties and stresses focus on correspondence establishment from 2D touchless fingerprint images and finger shape model estimation among four steps for 3D reconstruction. Specific to the characteristics of low ridge-valley contrast of touchless fingerprint images, we improved fingerprint enhancement method. So as to extract more robust fingerprint features. Then, three frequently used features, scale invariant feature transformation (SIFT) feature, ridge feature and minutia, are considered for correspondence establishment. Correspondences are finally established by adopting hierarchical fingerprint matching approaches. Finger shape model in our chapter is estimated by analyzing 3D point cloud finger data collected by one camera and a projector using the SLI method. We found binary quadratic function is more suitable for finger shape model compared with another mixed model we used in this chapter. According to reconstruct a standard cylinder object, we proved that it is reasonable and feasible of the adopted methodology of reconstruction technique and our capturing device. The comparison and analysis of 3D fingerprint reconstruction results from different fingerprint feature correspondences illustrates best reconstruction results can be generated if all three features of correspondences are used.

References

Bouguet JY (2010) Camera Calibration Toolbox for Matlab http://www.vision.caltech.edu/bouguetj/calib_doc/index.html

Choi H, Choi K, Kim J (2010) Mosaicing touchless and mirror-reflected fingerprint images. IEEE Trans Inf Forensics and Secur 5(1):52–61. doi:10.1109/TIFS.2009.2038758

Choi K, Choi H, Lee S, and Kim J (2007) Fingerprint image mosaicking by recursive ridge mapping, special issue on recent advances in biometrics systems. IEEE Trans Syst, Man, Cybern B 1191–1203. doi:10.1109/TSMCB.2007.907038

Feng J (2008) Combining minutiae descriptors for fingerprint matching. Pattern Recogn 342–352. doi:10.1016/j.patcog.2007.04.016

Hong L, Wan Y, and Jain A K (1998) Fingerprint image enhancement: Algorithms and performance evaluation. IEEE Trans Pattern Anal Mach Intell 777–789. doi:10.1109/34.709565

Jain A, Ross A (2002) Fingerprint mosaicking. In: Proceedings of IEEE international conference on acoustics, speech, and signal processing (ICASSP) 13–17. doi:10.1109/ICASSP.2002.5745550

Jain AK, Hong L, Bolle RM (1997) On-line fingerprint verification. IEEE Trans Pattern Anal Mach Intell 19(4):302–314. doi:http://doi.ieeecomputersociety.org/10.1109/34.587996

Jain AK, Chen Y, Demirkus M (2007) Pores and ridges: high resolution fingerprint matching using level 3 features. IEEE Trans Pattern Anal Mach Intell 29(1):15–27. doi:http://doi.ieeecomputersociety.org/10.1109/TPAMI.2007.17

Kumar A, Zhou Y B (2011) Contactless fingerprint identification using level zero features. Proc CVPRW 2011 121–126. doi:10.1109/CVPRW.2011.5981823

Lowe DG (2004) Distinctive image features from scale-invariant keypoints. Int J Comput Vis 60(2):91–110. doi:10.1023/B:VISI.0000029664.99615.94

Malathi S, Meena C (2010) Partial fingerprint matching based on SIFT features. Int J Comput Sci Eng 2(4):1411–1414. doi:http://www.doaj.org/doaj?func=openurl&genre=article&issn=09753397&date=2010&volume=2&issue=4&spage=1411

Maltoni D, Maio D, Jain A, Prabhakar S (2009) Handbook of fingerprint recognition. Springer, New York. ISBN 1848822537

Park U, Pankanti S, Jain A K (2008) Fingerprint verification using SIFT features. Proc SPIE6944, 69440K–69440K-9. doi:10.1117/12.778804

Shah S, Ross A, Shah J, and Crihalmeanu S (2005) Fingerprint mosaicing using thin plate splines In: The biometric consortium conference

Zhang D, Liu F, Zhao Q, Lu G, Luo N (2011) Selecting a Reference High Resolution for
    Fingerprint Recognition Using Minutiae and Pores. IEEE Trans Instrum Measur 863–871.
    doi:10.1109/TIM.2010.2062610
Zhang Z (2000) A flexible new technique for camera calibration. IEEE Trans Pattern Anal Mach
    Intel 22(11):1330–1334. doi:10.1109/34.888718

Chapter 12
3D Fingerprint Identification System

Abstract Human finger is a three-dimensional object. More real and more fingerprint features will be provided if 3D fingerprint images are available. This chapter thus explores 3D fingerprint features and their applications for personal identification. We define the 3D finger structural features, such as curve-skeleton, sectional curvatures as Level Zero Fingerprint Features in this chapter and investigate their distinctiveness for personal identification. These features are also used to assist fingerprint matching and make contribution to fingerprint recognition by combining with 2D fingerprint features. A series of experiments is conducted to evaluate 3D fingerprint recognition technique based on our established database with 541 fingers. Results show that an accuracy of 84.7 % can be achieved when using 3D curve-skeleton for recognition. The sectional curvatures can be used for human gender classification and an accuracy of 81 % is obtained in our database. An EER of 3.4 % is realized by including Level Zero Features into fingerprint recognition which demonstrates the effectiveness of 3D fingerprint recognition.

Keywords 3D fingerprint recognition • Level zero features • Curve skeleton • Sectional curvatures • Gender classification

12.1 Introduction

Automated personal identification based on fingerprints has been studied for centuries. Quite effective Automated Fingerprint Identification Systems (AFISs) are available with the rapid development of fingerprint acquisition devices and the advent of lots of advanced fingerprint recognition algorithms. However, they are almost based on 2D fingerprint features, even though the fact is that human finger is a 3D object. There are distortions and deformations introduced and 3D information lost when 2D fingerprint images used, which cannot perfectly meet peoples' demands in accuracy, computational complexity and user-friendly. Develop user-friendly AFIS with high precision and high efficiency is still an open issue in fingerprint recognition domain.

D. Zhang and G. Lu, *3D Biometrics*, DOI: 10.1007/978-1-4614-7400-5_12,   217
© Springer Science+Business Media New York 2013

With the expansion of acquisition technology, 3D biometrics authentication techniques come into researchers' view in recent years, such as 3D face (Samir et al. 2006; Lu et al. 2006), 3D ear (Yan et al. 2005; Chen et al. 2007; Yan et al. 2007) and 3D palmprint recognition (Zhang et al. 2008; Li et al. 2010; Zhang et al. 2009; Li et al. 2009). For 3D fingerprints, even though there are some works about 3D fingerprint image acquisition and processing (Parziale et al. 2006; Wang et al. 2010), they did not investigate the utility of 3D fingerprint features and did not report any experimental results of personal identification using the acquired biometrics information. This has motivated us to explore the utility of 3D fingerprint features and the possibility of combining them with 2D features for fingerprint recognition. The contributions of this chapter includes: (1) this chapter, for the first time, investigates features on 3D fingerprint images, including Level Zero Features for fingerprint recognition, the corresponding feature extraction and matching method is proposed. More specifically, the 3D finger structural features, such as curve-skeleton, overall surface curvature features are firstly defined as Level Zero Features, and then extracted by model fitting method, Iterative Closest Point (ICP) in 3D space is finally adopted to matching; (2) By analyzing their distinctiveness, Level Zero features are used for different applications. We found curve-skeleton are suitable for assisting fingerprint recognition while overall surface can be used for gender classification; (3) Fusion strategy is employed to combine 2D and 3D fingerprint matching result to figure out the effectiveness of improving recognition accuracy by including 3D fingerprint features.

12.2 Definition of Level Zero Features in 3D Fingerprint Images

In general, fingerprints are distinguished by their features. Fingerprint features in 2D images are classified into three levels (Ashbaugh 1999), as shown in. Level 1 features are the macro details of fingerprints such as singular points and global ridge patterns, such as deltas and cores. They are not very distinctive and are thus mainly used for fingerprint classification rather than recognition. The level 2 features primarily refer to the Galton features or minutiae, namely ridge endings and bifurcations. Level 2 features are the most distinctive and stable features, which are used in almost all automated fingerprint recognition systems (AFRS) (Ashbaugh 1999; Maltoni et al. 2009; Ratha et al. 2004). Level 3 features (red circles) are often defined as the dimensional attributes of the ridges and include sweat pores, ridge contours, and ridge edge features, all of which provide quantitative data supporting more accurate and robust fingerprint recognition. However, see fingerprint image in 3D space, as shown in Fig. 12.1, we can got that the above defined fingerprint features spread over different scales of depth. For example, core points locate in the center part of the finger with almost the highest depth value. Level 2 and Level 3 features which closely related with the distribution of ridges actually possess more attributes in 3D space (e.g. depth value, ridge orientation along depth direction). Thus, in this chapter, we defined the structural

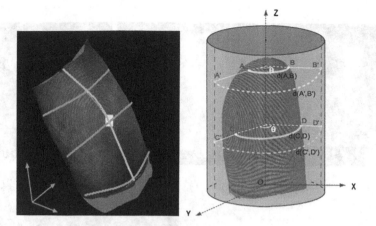

Fig. 12.1 Fingerprint image in 3D space

information in 3D fingerprint images as Level Zero Features. They provide information of overall structure of humans' finger, such as the curve-skeleton (Cornea et al. 2007) and overall curvature. The curve-skeleton feature depicts the thinned contour of finger shape, as shown in Fig. 12.1 (green and red lines). The overall curvature describes the maximal horizontal and vertical curvature of finger since it is curved shape of human finger.

12.3 3D Fingerprint Image Analysis

12.3.1 Source of the 3D Fingerprint Image

Currently, there are three frequently-used 3D imaging techniques, namely multi-view reconstruction (Parziale et al. 2006; Hartley 2000; Hernandez et al. 2008), laser scanning (Blais 1988), and structured light scanning (Wang 2010; Stockman et al. 1988; Hu and Stockman 1989). Among them, the multi-view reconstruction technique has the advantage of low-cost but disadvantage of low accuracy. Laser scanning normally achieves high resolution 3D images but cost expensive and long collecting time. It is also very sensitive to the status (wet or dry) of object. Structured light imaging is a 3D scanning technique which has moderate accuracy and cost, but also takes long time to collect 3D data. It has the drawback of instability to move which means one should keep still when projecting some structured light patterns to human finger. Considering the cost, usability, as well as just structural information needed, we choose to use multi-view reconstruction technique to acquire the 3D fingerprint image. The reconstruction results from a touchless multi-view fingerprint imaging device in our own designing and the 3D reconstruction techniques we proposed is introduced in Chap. 11. Figure 12.2 shows an example of captured images and the reconstructed 3D fingerprint image.

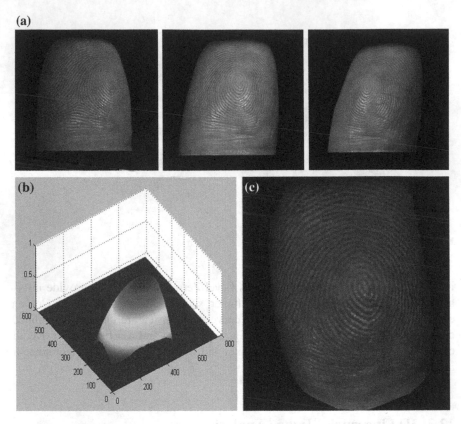

Fig. 12.2 Captured fingerprint images by our designed touchless multi-view imaging device and 3D finger shape: **a** images of a finger captured by our device (*left, frontal, right*), **b** reconstructed 3D finger shape, **c** 3D fingerprint image

12.3.2 Level Zero Feature Extraction

Since our 3D fingerprint image is reconstructed from multi-view fingerprint images, there is a one-to-one correspondence between the 3D points and the 2D fingerprint image pixels. Preprocessing such as ROI extraction and pose correction can be done in 2D fingerprint images, and implement into 3D situation. The iterative thresholding segmentation method introduced in Chap. 10 is used in this chapter to extract ROI (see Fig. 12.4b). Since the uncontrolling of fingerprint image collection (tilted of finger position, see Fig. 12.3a), pose correction is necessary. We accomplished it by simple rotating the original image as follows: (1) Scan each line of ROI horizontally and find the center point (green line in Fig. 12.3b); (2) Fit such center points by a line (red line in Fig. 12.3b); (3) Calculate the angle between the fitted center line and vertical axis (θ shown in Fig. 12.3b); (4) Rotate the image anticlockwise by θ. Figure 12.3c shows the final correct 2D fingerprint image and Fig. 12.3e shows the correct 3D finger shape of original 3D shape of Fig. 12.3d.

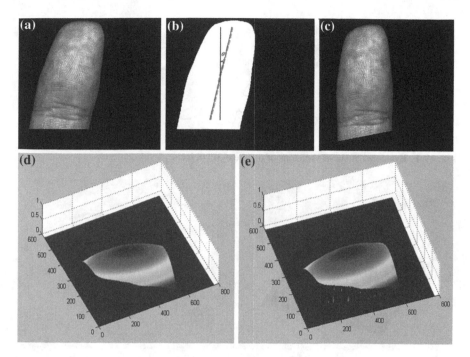

Fig. 12.3 Position correction: **a** original tilted fingerprint image, **b** ROI of (**a**), **c** fingerprint image after pose correction, **d** original 3D finger shape, **e** corrected 3D finger shape

Given a corrrected 3D fingerprint image, stable and unique features are expected to be extracted for the following pattern matching and recognition. 3D depth information reflects the overall structure of human finger. However, there are many invalid points in the whole 3D finger shape due to the structure of human finger. Wrinkles and scars in finger also affect local structure of finger shape. Thus we proposed to extract curve-skeleton of finger shape. As shown in Fig. 12.4, Different 3D objects almost fully represented by their curve-skeletons.

Since 3D finger shape model is close to binary quadratic function, profile of horizontal section can be fitted by parabola and reflects the changes of finger width, while vertical profile depicts variation tendency of depth from finger tip to distal interphalangeal crease. The curve-skeleton of 3D fingerprint image consists of representative vertical and horizontal lines. Since each horizontal profile is parabola-like shape, we extracted the extreme value of each fitted parabola line to form the representative vertical line (blue line in Fig. 12.5a). Three representative horizontal lines are selected at a certain step length (100). The distal interphalangeal crease is chosen as the base line (green line in Fig. 12.5a). Figure 12.5b then shows the curve-skeleton we extracted from 3D finger depth map. For overall curvatures, they can be easily calculated since our 3D finger shape is reconstructed by model fitting. The coefficients of the binary quadratic function control the maximal horizontal and vertical curvatures of 3D finger, namely the parameters of A and B in Eq. (12.1). Thus, these two coefficients of the binary quadratic function are maintained to represent the maximal horizontal and vertical curvatures, namely the defined overall curvatures.

Fig. 12.4 Examples of curve-skeletons of different 3D objects

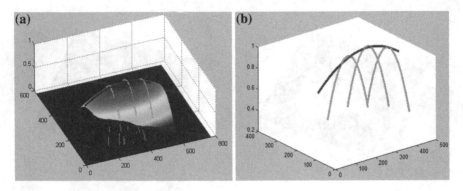

Fig. 12.5 Examples of curve-skeleton for 3D finger: **a** 3D finger shape, **b** extracted curve-skeleton

$$f(x, y) = Ax^2 + By^2 + Cxy + Dx + Ey + F \tag{12.1}$$

From Fig. 12.5b, we can see that curve-skeleton consists of several 3D lines. Intuitively, the iterative closest point (ICP) algorithm is suitable to solve such matching problem. ICP method (Besl et al. 1992) is widely used in many 3D object recognition systems for matching. In this chapter, we slightly modified the ICP method to meansure the distances between two sets of points. The algorithm is given below and Fig. 12.6 shows an example of matching two curve skeletons by our modified ICP method. ICP algorithm:

1. Input: Medel point set: D_1; Test point set D_2;
2. Parameters initialization: stop criterion for distance $T_d = 0.1$; intial rotation matrix $R_0 = I$; initaltranslation vector $T_0 = [0\ 0\ 0]^T$;
3. While (new correspondences set found between D_1 and D_2))

   {[corr, D_i] = dsearchn(D_1, D_2);

   $K_i = D_i > T_d$;
   Discard corr(K_i);
   Update R_i, T_i;
   $D_2 = R_i * D_2 + T_i$;}

4.   Output: distance vector D, registered D2, rigid transform paramters: R and T.

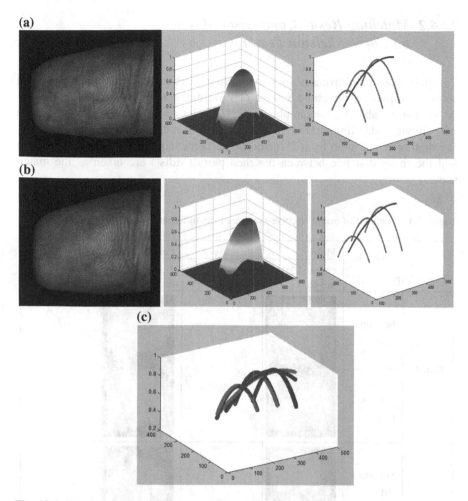

Fig. 12.6 Example of curve-skeleton matching by icp method: **a** the model 2D fingerprint image, 3D finger shape, and extracted curve-skeleton feature, **b** the test 2D fingerprint image, 3D finger shape, and extracted curve-skeleton feature, **c** matching result by icp method

12.4 Experimental Results and Analysis

12.4.1 Dataset

It is notable that our 3D fingerprint images consist of the reconstruction results from a touchless multi-view fingerprint imaging device designed by our group. The 3D reconstruction techniques we proposed are introduced in Chap. 11. Our experiments are then implemented on our reconstructed 3D fingerprint dataset with 541 fingers, each 2 pictures captured at separate sessions apart from one week to months. These fingers come from volunteers with 223 female fingers and 318 male fingers, all five human fingers are included.

12.4.2 Matching Result Comparison Based on Curve-Skeleton Features

To study the distinctiveness of curve-skeleton features of human fingers, we show examples of matching results of different gender and different fingers. As shown in Table 12.1, examples of curve-skeletons from a female and a male with thumb, index finger and little finger captured at different sessions are given. We then matched them by ICP method. The percentage of matched points (Pm) and the mean distance between matched pairs (Mdist) are taken as the match score.

Table 12.1 Examples of extracted curve-skeletons from different gender and different fingers

Gender / Finger Type		Male		Female	
		Orignial 2D image	Curve-skeleton	Orignial 2D image	Curve-skeleton
Thumb	Session 1 (a1)				
	Session 2 (a2)				
Index Finger	Session 1 (b1)				
	Session 2 (b2)				
Little Finger	Session 1 (c1)				
	Session 2 (c2)				

We firstly matched the curve-skeletons from the same finger but captured at different time, as listed in Table 12.2. Results show that the mean distance between matched pairs are smaller than 1 and the percentage of matched points are larger than 70 %. Figure 12.7 also shows the matching results of different gender and finger types, the match scores are listed in Table 12.3. The results show that big difference existed between different fingers and different genders in curve-skeleton, since such feature reflects the ridge width feature of human finger and curvatures are different for human finger from finger tip to the distal interphalangeal crease.

Fingerprint identification experiment based on curve-skeletons is then implemented on our established database. Figure 12.8 shows the ROCs of different match score indexes. The EERs were obtained from 541 genuine scores and 292,140 imposter scores (generated from 541 fingers, 2 pictures of each finger). From the results, we can see that an EER of around 15 % can be obtained when matching 3D fingerprint curve-skeleton features by simple ICP algorithm. The index of mean distance between matched pairs is better than the percentage of matched points. Curve-skeleton feature of 3D fingerprint image can be used to distinguish different fingers even though it is not as accurate as other higher level fingerprint features.

Table 12.2 Matching result comparison of curve-skeletons from the same finger but different session

Finger Type / Gender	Thumb (a1)—(a2)	Index Finger (b1)—(b2)	Little Finger (c1)—(c2)
Male	Pm=74%; Mdist=0.20	Pm=93%; Mdist=0.39	Pm=79%; Mdist=0.25
Female	Pm=94%; Mdist=0.72	Pm=97%; Mdist=0.09	Pm=90%; Mdist=0.32

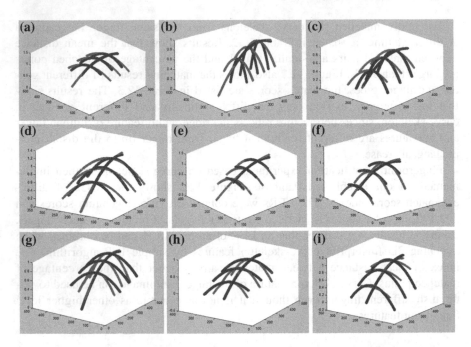

Fig. 12.7 Example of matching results of curve-skeletons from different gender and finger types: **a** matching result of [(male, thumb)–(male, index finger)] in Table 12.1, **b** matching result of [(male, thumb)–(male, little finger)] in Table 12.1, **c** matching result of [(male, index finger)–(male, little finger)] in Table 12.1, **d** matching result of [(female, thumb)–(female, index finger)] in Table 12.1, **e** matching result of [(female, thumb)–(female, little finger)] in Table 12.1, **f** matching result of [(female, index finger)–(female, little finger)] in Table 12.1, **g** matching result of [(male, thumb)–(female, thumb)] in Table 12.1, **h** matching result of [(male, index finger)–(female, index finger)] in Table 12.1, **i** matching result of [(male, little finger)–(female, little finger)] in Table 12.1

Table 12.3 Match scores corresponding to Fig. 12.7

Corresponding labels in Fig. 12.7 Match score index	a	b	c	d	e	f	g	h	i
Pm (%)	57	38	53	55	45	62	50	53	57
Mdist	8.3	13.7	2.9	6.8	14.8	3.1	4.0	4.1	4.6

12.4.3 Gender Classification Based on Overall Curvature Feature

Since our 3D fingerprint images are generated by reconstruction where binary quadratic function is taken as the finger shape model, two parameters are used to depict the overall finger shape curvature. Figure 12.9 shows the values of maximal horizontal curvature feature and maximal vertical curvature feature in our database. We found both of these curvature features are very small. It cannot be used for personal identification. Thanks to the composition of database of different

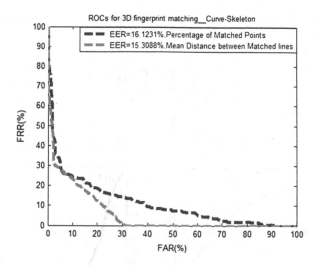

Fig. 12.8 ROCs for 3D fingerprint matching by ICP with curve-skeleton feature

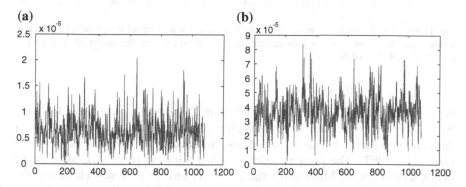

Fig. 12.9 Values of overall curvature features on our database: **a** horizontal curvature feature, **b** vertical curvature feature

gender, we investigated whether this feature is useful for gender classification. We then plot the distribution maps of norm curvature features separated by gender, as shown in Fig. 12.10. The ROCs are also shown in Fig. 12.10. From the figure, we found that the vertical curvature features can reach classification accuracy at 81 %, while horizontal curvature feature disabled to classify genders. It stated that there is little difference in horizontal profile for different fingers no matter male or female but differ in their vertical profile.

12.4.4 Fusion Results of 2D and 3D Fingerprint Features

Since both 2D fingerprint features and 3D structural features are provided simultaneously by 3D fingerprint images, we aim to study whether improved

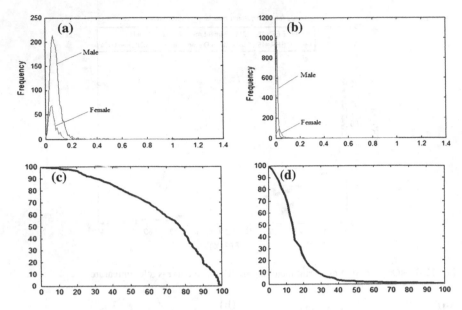

Fig. 12.10 Overall curvature features for gender classification: **a** distribution map of horizontal curvature feature for different gender, **b** distribution map of vertical curvature feature for different gender, **c** ROC of (**a**), **d** ROC of (**b**)

performance can be achieved by combining 2D and 3D fingerprint features. For 2D fingerprint features, we selected minutiae due to their distinctiveness and popularity. It was extracted and matched by the method proposed in (Jain et al. 1997). The percentage of matched minutiae pairs was taken as the match score (MS_{2D}). Meanwhile, the curve-skeleton feature was chosen as the 3D structural fingerprint feature and mean distance between matched pairs was taken as the match score (MS_{3D}). A simple adaptive weighted sum rule is used to combine the 2D and 3D matching scores. The combined score can be expressed as

$$MS_{2D+3D} = w/MS_{3D} + (1 - w) \times MS_{2D} \quad w \in [0,1] . \tag{12.2}$$

The weight w is adaptively tuned to provide the best verification results at step length of 0.01.

Figure 12.11 shows the ROCs achieved by using minutiae and curve-skeleton separately, as well as their combination. It is notable that minutiae clearly outperforms curve-skeleton in terms of accuracy. However, best result is achieved when combining minutiae and curve-skeleton feature where an EER of 3.4 % is obtained. This experiment fully demonstrates that higher accuracy can be achieved if 3D fingerprint images used compared with 2D fingerprint recognition.

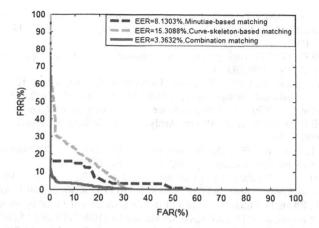

Fig. 12.11 ROCs for fingerprint matching by different fingerprint features

12.5 Summary

This chapter has proposed a 3D fingerprint recognition technique. Thanks to the available of 3D fingerprint images, more features can be extracted. A coarser level of fingerprint features than level 1 features—Level Zero Features which refers to finger structural features are firstly defined in this chapter. These features are then used for assisting fingerprint identification and gender classification. Experimental results show that an accuracy of 84.7 % can be achieved when using 3D curve-skeleton for recognition. The sectional curvatures can be used for human gender classification and an accuracy of 81 % is obtained in our database. An EER of 3.4 % is realized by including Level Zero Features into fingerprint recognition which demonstrates the effectiveness of 3D fingerprint recognition. Simple feature extraction and matching algorithm are used in this chapter. We believe that higher accuracy can be achieved if more advanced feature extraction and matching methods are proposed in the future. Discovering the relationship between different levels of fingerprint features and designing more powerful fusion strategy will further improve 3D fingerprint recognition performance.

References

Ashbaugh DR (1999) Quantitative-qualitative friction ridge analysis: an introduction to basic and advanced ridgeology. CRC Press, Boca Raton. ISBN 0849370078

Besl PJ, McKay ND (1992) A method for registration of 3-D shapes. IEEE Transactions on Pattern Analysis and Machine Intelligence, pp 239–256. doi:10.1109/34.121791

Blais F, Rious M, Beraldin JA (1988) Practical considerations for a design of a high precision 3-D laser scanner system.In: Proceedings of SPIE, pp 225–246. doi:10.1117/12.947787

Chen H, Bhanu B (2007) Human ear recognition in 3D. IEEE Transactions on Pattern Analysis and Machine Intelligence, pp 718–737. doi:10.1109/TPAMI.2007.1005

Cornea ND, Silver D,Min P (2007) Curve-skeleton properties, applications, and algorithms. IEEE Transactions on Visualization and Computer Graphics, pp 530–548. doi:10.1109/T VCG.2007.1002

Hartley R (2000) Multiple view geometry in computer vision. Cambridge University Press, Cambridge (ASIN: B008Q3O7FC)

Hernandez C, Vogiatzis G, Cipolla R (2008) Multiview photometric stereo. IEEE Transactions on Pattern Analysis and Machine Intelligence, pp 548–554. doi:10.1109/TPAMI.2007.70820

Hu G, Stockman G (1989) 3-D surface solution using structured light and constraint propagation. IEEE Transactions on Pattern Analysis and Machine Intelligence, pp 390–402. doi:10.1109/34.19035

Jain A, Hong L, Bolle R (1997) On-line fingerprint verification. IEEE Transactions on Pattern Analysis and Machine Intelligence, pp 302–314. doi:10.1109/34.587996

Li W, Zhang D, Zhang L (2009) Three Dimensional Palmprint Recognition. IEEE International Conference on Systems, Man, and Cybernetics. doi:10.1109/ICSMC.2009.5346053

Li W, Zhang L, Zhang D, Lu G, Yan J (2010) Efficient joint 2d and 3d palmprint matching with alignment refinement. In: Proceedings of CVPR. doi:10.1109/CVPR.2010.5540134

Lu X, Jain AK, Colbry D (2006) Matching 2.5D face scans to 3D models. IEEE Transactions on Pattern Analysis and Machine Intelligence, pp 31–43. doi:10.1109/TPAMI.2006.15

Maltoni D, Maio D, Jain A, Prabhakar S (2009) Handbook of fingerprint recognition. Springer, New York. ISBN 1848822537

Parziale G, Diaz-Santana E (2006) The surround imager: a multi-camera touchless device to acquire 3D rolled-equivalent fingerprints. In: Proceedings of International Conference on Biometrics (ICB), Hong Kong, China, pp 244–250. doi:10.1007/11608288_33

Ratha N, Bolle R (2004) Automatic fingerptrint recognitiion systems. Springer, New York

Samir C, Srivastava A, Daoudi M (2006) Three-dimensional face recognition using shapes of facial curves. IEEE Transactions on Pattern Analysis and Machine Intelligence, pp 858–1863. doi:10.1109/TPAMI.2006.235

Stockman GC, Chen SW, Hu G, Shrikhande N (1988) Sensing and recognition of rigid objects using structured light. IEEE Control Syst Mag 14–22. doi:10.1109/37.472

Wang Y, Hassebrook LG, Lau DL (2010) Data acquisition and processing of 3-d fingerprints. IEEE Transactions on Information Forensics and Security, pp 750–760. doi:10.1109/T IFS.2010.2062177

Yan P, Bowyer KW (2005) Multi-biometrics 2D and 3D ear recognition. In: Proceedings of AVBPA, pp 503–512. doi:10.1007/11527923_52

Yan P, Bowyer KW (2007) Biometric recognition using 3D ear shape. IEEE Transactions on Pattern Analysis and Machine Intelligence, pp 297–1308. doi:10.1109/TPAMI.2007.1067

Zhang D, Lu G, Li W, Zhang L, Luo N (2008) Three dimensional palmprint recognition using structured light imaging, 2nd ieee international conference on biometrics: theory, applications and systems, BTAS, pp 1–6. doi:10.1109/BTAS.2008.4699346

Zhang D, Lu G, Li W, Zhang L, Luo N (2009) palmprint recognition Using 3-D information. IEEE Transactions on Systems, Man, and Cybernetics, Part C: Applications and Reviews, pp 505–519. doi:10.1109/TSMCC.2009.2020790

Part V
3D Face Verification by Tof Method

Chapter 13
The Principle of 3D Camera Imaging

Abstract Acquiring 3D geometric information from real environments is an essential task for many applications in computer graphics and computer vision. It is obvious that lots of application can be benefit from simple and accurate devices for real-time range image acquisition. However, even for static scenes there is no low-price off-the shelf system can provide full-range, high resolution distance information in real time. Time-of-Flight (ToF) technology, based on measuring the time that light emitted by an illumination unit requires to travel to an object and back to a detector, is the basis for the development of new range-sensing devices, so-called ToF cameras, which are realized in standard CMOS or CCD technology. A ToF camera will be introduced in this chapter for acquiring 3D face images.

Keywords Time-of-Flight (ToF) technology • 3D face image

13.1 Introduction of Time-of-Flight Camera

Acquiring 3D geometric information from real environments is an essential task for many applications in computer graphics and computer vision. It is obvious that lots of application can be benefit from simple and accurate devices for real-time range image acquisition. However, even for static scenes there is no low-price off-the shelf system can provide full-range, high resolution distance information in real time. For example, the laser triangulation techniques, which sample a scene row by row with a laser device, are rather time-consuming and impracticable for most application in real world.

Time-of-Flight (ToF) technology, based on measuring the time that light emitted by an illumination unit requires to travel to an object and back to a detector, is used in Light Detection and range scanners for high-precision distance measurements. Recently, this principle has been the basis for the development of new range-sensing devices, so-called ToF cameras, which are realized in standard CMOS or CCD technology. There are two main approaches currently employed in

D. Zhang and G. Lu, *3D Biometrics*, DOI: 10.1007/978-1-4614-7400-5_13,
© Springer Science+Business Media New York 2013

Fig. 13.1 ToF camera

ToF technology. The first one utilizes modulated, incoherent light, and is based on a phase measurement. The second one is based on an optical shutter technology.

This ToF principle is used by various manufactures, e.g. PMDTec/ifm electronics (www.pmdtec.com, Fig. 13.1, left), MESA Imaging (www.mesa-imaging.ch, Fig. 13.1 right). And it has been used in substantial number of application areas such as automated production or automotive applications.

13.2 System Principle and Architecture of ToF

The intensity modulation principle (see Fig. 13.2) is based on the on-chip correlation (or mixing) of the incident optical signal s, coming from a modulated NIR illumination and reflected by the scene, with its reference signal g, possibly with an internal phase offset τ,

$$C(\tau) = s \otimes g = \lim_{T \to \infty} \int_{-T/2}^{T/2} s(t) \bullet g(t + \tau)dt. \tag{13.1}$$

For a sinusoidal signal, e.g.

$$g(t) = \cos(2\pi f_m t), \ s(t) = b + a\cos(2\pi f_m t + \phi), \tag{13.2}$$

where f_m is the modulation frequency, a is the amplitude of the incident optical signal, b is the correlation bias and ϕ is the phase offset corresponding to the object distance. So, we can get the follow result

$$C(\tau) = \frac{a}{2}\cos(f_m \tau + \phi) + b. \tag{13.3}$$

The demodulation of the correlation function c is done using samples of the correlation function c obtained by four sequential phase images with different phase offset τ:

$$A_i = C(i * \frac{\pi}{2}), i = 0, \ldots, 3; \ a = \frac{\sqrt{(A_3 - A_1)^2 + (A_0 - A_2)^2}}{2},$$

Fig. 13.2 The ToF phase-measurement principle

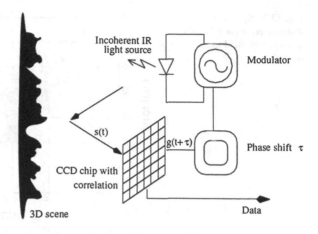

$$\phi = \arctan 2(A_3 - A_1, A_0 - A_2), \; I = \frac{A_0 + A_1 + A_2 + A_3}{4}, \qquad (13.4)$$

where I is the intensity of the incident NIR light. Now, from ϕ one can easily computer the object distance $d = \frac{c}{4\pi f_m}\phi$, where c is the speed of light. And most of the current cameras support Suppression of Background Intensity (SBI), which facilitates outdoor applications. If the sensor is equipped with SBI, the intensity I mainly reflects the incident active light.

Using ToF cameras based on the intensity modulation approach involves major sensor specific challenges:

1. Low Resolution: Current cameras have a resolution between $64 * 48$ and $200 * 200$. This resolution is rather small in comparison to standard RGB- or grayscale-cameras.
2. Systematic Distance Error: Since the theoretically required sinusoidal signal is not achievable in practice, the measured depth does not reflect the true distance but contains a systematic error, also called "wiggling" (see Fig. 13.3).
3. Intensity-related Distance Error: Additionally, the measured distance is influenced by the total amount of incident light. This fact results from different physical effects in the ToF camera, both the semiconductor detector and the camera electronics. However, this is not a generic ToF problem and some manufactures seem to have found solutions to this problem (see Fig. 13.4).

Regarding the systematic distance error, several approaches have been proposed. A very comprehensive study of systematic error of various ToF cameras has been carried out in Rapp (2007). One major result of their study is that systematic error behaves quite similarly for different camera types. Differences appear in the near range. By controlling the shutter time, the depth data can be optimized.

The noise level of the distance measurement is mainly affected by the amount of incident active light. In literature (May et al. 2006), they use the median filter, according to the depth value of the neighbourhood, to improve the accuracy of

Fig. 13.3 Systematic
distance error for all pixels
in *grey*, and fitted mean
deviation in *black*

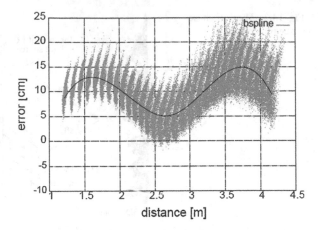

Fig. 13.4 Intensity-related
distance error when sensing
a planar object with varying
reflectivity

depth data; in (Fuchs and Hirzinger 2006), an error model combines the distance and the amplitude information was adopted to correct the deviation of the depth. Their error model consists of several splines; each one has different coefficients, which is estimated by their different location.

Optical Shutter Approach (see Fig. 13.5)—this alternative ToF principle is based on the indirect measurement of the time of flight using a fast shutter technique, realized, for example, in cameras from 3DV systems. The basic concept uses a short NIR light pulse $[t_{start}, t_{stop}]$, which represents a depth range of interest. We can see from Fig. 13.5, a "light wall" is emitted from the camera (top-left) and reflected by the object (top-right). Gating the reflected optical signal yields distance related portions of the "light wall", which are measured in the CCD pixels.

A shutter in front of a standard CCD camera cuts the front portion of the optical signal at the gating time $t_{gate} = t_{start} + \Delta t$. The resulting intensity I_{front} is proportional to the distance to the corresponding segment of the object's surface. This

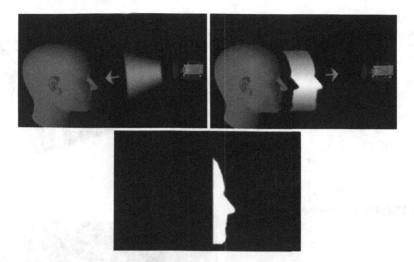

Fig. 13.5 The ToF shutter principle

normalizes the object's reflectivity as well as the attenuation of the active light due to the object's distance. The distance measurement relates to the "distance time range" and the relation $I_{\text{front}}/I_{\text{total}}$ of the total reflected signal and the front cut, i.e.:

$$d = (1 - \alpha)d_{\min} + \alpha d_{\max}, \alpha = I_{gate}/I_{total}$$

$$d_{\min} = c * t_{\min}, d_{\max} = c * t_{\max}, c \approx 3 * 10^8 \text{ m/s.} \qquad (13.5)$$

As we can see that distance beyond the range from $d_{\min}$ to $d_{\max}$ can't be measured in a single exposure.

PMD CamCube 3.0 (PMD Tutorial[1]) (see Fig. 13.6), the camera we used in our system. It adopts the intensity modulation principle. We can see its key features at a glance: 200 × 200 pixels; ranges of 7.5 m; record surroundings at the high rate of 25 frames per second; suitable for indoor and outdoor environments, thanks to SBI module; simultaneous capture of grayscale images and distance information; high-speed interface providing raw data for image processing; API and MATLAB interface for Linux and Windows.

Figure 13.7 demonstrates the setup of our system, where ToF depth camera is set in a "black box". For this "black box", the material of its front side is carefully selected so that it allows photon emitted by ToF depth camera to freely

---

[1] http://www.pmdtec.com/fileadmin/pmdtec/downloads/documentation/datenblatt_camcube3.pdf

Fig. 13.6 PMD CamCube,
a camera unit, **b** illumination
unit

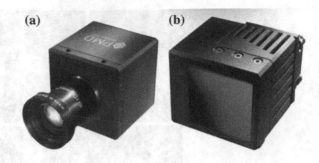

Fig. 13.7 Our system setup

pass through it. Obviously, our setup is robust to illumination changes, which are thought to be great challenges in traditional 2D face recognition.

We can get three types of data from this camera at same time. The most important image is of course the distance image (see Fig. 13.8a). In each pixel, it provides us with a distance from the camera to the observed object. The amplitudes (see Fig. 13.8b) provide additional useful information: The higher the amplitude value of a pixel, the more reliable is its corresponding distance value. If the camera observes a scene with good reflectivity, the amplitude values will be high. If the object is beyond the measurement range, the amplitudes will be close to zero. Finally, the intensity image (see Fig. 13.8c) is similar to a simple grayscale image from a traditional 2D camera: the more light reaches a pixel, the higher is its intensity value.

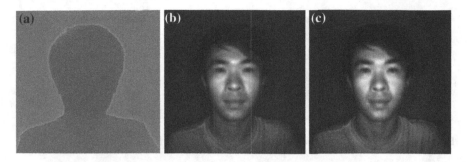

Fig. 13.8 Data from PMD CamCube 3.0: **a** depth map, **b** amplitude map, **c** intensity map

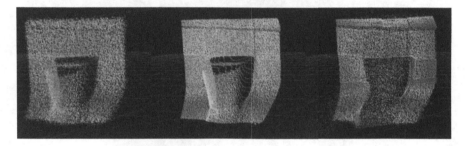

Fig. 13.9 Different integration time of same scene

The most important parameter about the camera is the integration time. But unfortunately, a general suggestion for an integration time cannot be made, because the data acquisition depends on too many factors.

In Fig. 13.9, from left to right, the integration times is 140, 1,400 and 14,000 μs. Note the low signal strength on the left and the saturation on the right side due to unsuitable integration times. When choosing the right integration time, we have to take signal strength and saturation into account.

Currently, there are some hot research areas about the application using ToF camera.

1.   Geometric feature extraction and 3D reconstruction

Since the ToF can capture the 3D data in real-time, the ToF camera can be used in either static or dynamic scenes. After the 3D data of the environment have been captured by the camera, the 3D data of whole scene can be reconstructed in a unified coordinate system (Huhle et al. 2008). Kim et al. (2009) proposed a multi-sensor fusion method. It uses several color cameras and ToF cameras, and combined all these data together. The result is illustrated in Fig. 13.10.

Also the resolution and data quality of the ToF camera is not good enough. How to use this kind of data to generate accurate 3D model? Lately Cui et al.

Fig. 13.10   Reconstruction of the 3D scene from multi-sensor

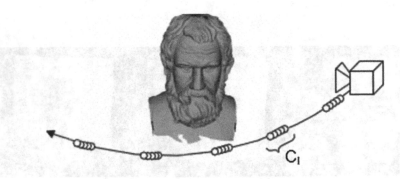

Fig. 13.11   The path of camera movement

(2010) has proposed a simple method to scan 3D object (see Fig. 13.11): they capture the data of 3D object while the camera move around it. And then, all the data are processed, combined and optimized.

There is another way for high quality 3D reconstruction—structure for motion, SFM (Bartczak et al. 2007; Koeser et al. 2007). The defect about SFM is the loss of measurement of scale. This problem can be solved by combined with a ToF camera, which provides the measurement of scale character (Streckel et al. 2007). In Tong et al. (2007), they have proposed a 3D reconstruction method using a single ToF camera, and they also applied this method to hair-style scan, see Fig. 13.12.

2.   Human–Computer interaction

Since the ToF camera can provide 3D information in real-time, it have lots of advantage compared with some traditional device. In (Xiang et al. 2011), Oggier uses it to track the movement of human hand, and develop an interactive touch system. Soutschek et al. (2008) proposed a similar scheme in the area of 3D medical visualization.

Zhu et al. (2009) proposed an auto-matting algorithm using ToF camera combined with high-resolution color camera. They use the depth image to generate the Tri-Map which usually needs user interaction. They calculate the depth map and alpha map in closed-form under the framework of Markov Random Field. The pipeline of this method is illustrated in Fig. 13.13.

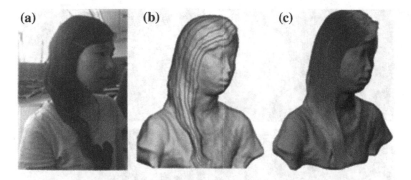

Fig. 13.12   Hair-style scan using a single ToF camera (Tong et al. 2007)

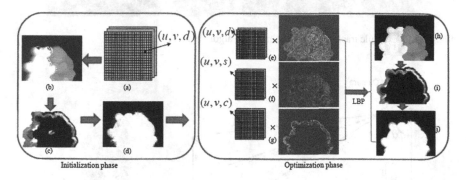

Fig. 13.13   The pipeline of auto-matting algorithm

Cai et al. (2010) have proposed a 3D face tracking algorithm using one single ToF camera. In particular, they adopt the ICP method and L1 normalization scheme to optimize the depth data for the 3D face tracking. You can find some result in Fig. 13.14.

Depth cameras have also been successfully used in the real-time marker-less three dimensional interactions, by detecting hand gestures and movements (Penne et al. 2008). Holte et al. (2008) first used depth cameras for gesture recognition, where only range data are used. Motion is detected using band-pass filtered difference range images (Holte et al. 2008). Then, they extended this to full body gesture recognition using spherical harmonics.

3.   Human-Oriented Analysis

Some medical applications such as cancer treatment require a re-positioning of the patient to a previously defined position. Depth cameras have been used in such situation to solve the problem by segmenting the patient body and registering a rigid 3D-3D surface registration (Adelt et al. 2008). Also, in iris capturing scenario, it has been reported that (Huang et al. 2009), depth sensor was used in iris

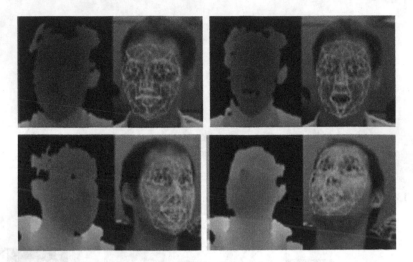

Fig. 13.14 Example tracking results using Cai's algorithm

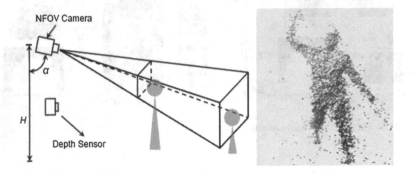

Fig. 13.15 *Left* The configuration of iris capturing system. *Right* The raw depth image

deblurring algorithm for less intrusive iris capture while improving the robustness and non-intrusiveness for iris capture.

Automatic detection and pose estimation of humans is an important task in Human Computer Interaction (HCI). Jain and Subramanian (2010) presented a model based approach for detecting and estimating human pose by fusing depth and RGB color data from monocular view. A further study was released by Ganapathi in Ganapathi et al. (2010) where they derive an efficient filtering algorithm for marker-less tracking human pose in real-time, using a stream of monocular depth images (see Fig. 13.15, right). The key idea lies in their approach is to combine an accurate generative model-which is achievable using programmable graphics hardware-with a discriminative model that feeds data driven evidence about body part locations. Since the accurate real-time tracking of humans and other articulated bodies is one that has enticed researchers for many years, their work opens a new door for the large number of useful applications.

13.3 Pre-Processing

We should do some pre-processing about the source data. This chapter presents the fundamentals of the technique we used in this step: Denoising, Face detection and Active shape model. Section 13.3.1 discusses the noise character of ToF data and how we remove the noise as much as possible. Sections 13.3.2 and 13.3.3 the face detection method and ASM are presented.

13.3.1 Denosing of Range Data

As we mentioned before, the data we capture from ToF sensor contain lots of noise. Since our device: PMD Cam-Cube 3.0, adopts the phase-shifting technique, the standard deviation σ of the range measurement is mutual to the amplitude A of the optical signal $\sigma = {}^{1}/_{A}$. From it, we know that the most popular 4-phase-shifting technique leads to estimation the angular component of a random variable with values in R^2; the mean length of this variable is given by the signal amplitude. We can assume that a uniform reflectivity of the facial surface. But it should be noted that the systematic errors are not considered. Cause this kind of errors are either difficult to control or can be eliminated after a sensor calibration. This error can't be considered as the noise.

In order to remove as many noises as possible, but still keep more detail, like edge, valley, etc.; we adopt the method proposed in (Thouis et al. 2003). While most previous work about mesh smoothing has favored diffusion-based iterative techniques for feature-preserving smoothing, they use a radically different approach, based on robust statistics and local first-order predictors of the surface. It is simple to implement and has better efficiency than the other methods (Fig. 13.16).

13.3.2 Face Detection

Although the resolution of our data is only 200 * 200, it still has lots of unrelated information in the data. So for each frame we get, we need to crop the face area. Since we only need the tackle the situation when there is a face exists in front of the camera. We need to detect the face existence before we carry on the feature extraction and verification process.

Many algorithms implement the face-detection task as a binary-classification task. That is, the content of a given part of an image is transformed into features, after which a classifier trained on example faces decides whether that particular region of the image is face or not. A window-sliding technique is often employed.

We adopt the Viola-Jones Object detection framework. It is the first object detection framework to provide competitive object detection rates in real-time.

Fig. 13.16 Denoising result

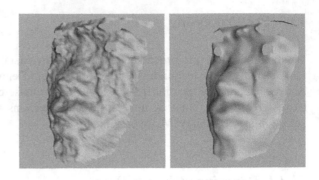

Fig. 13.17 Feature types

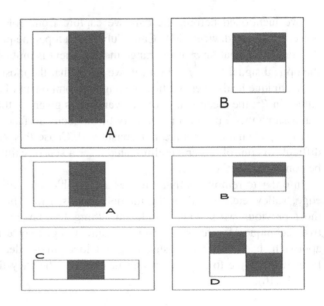

They use the Haar-like feature to represent the distinctive character of human face. Figure 13.17 illustrates four different types of features used in the framework. The value of any feature is always simply the sum of the pixels within clear rectangles subtracted from the sum of the pixels within shaded rectangles. An important technique, called the integral image, is used to fast calculate the feature in constant time.

No matter how simple the feature is, the number of total features is quite large. So it would be prohibitively expensive to evaluate them all. The object detection framework employs a variant of the learning algorithm AdaBoost to both select the most distinctive features and to train classifiers that use them.

If we train the classifier use the original recursive process, we can get a strong classifier, but it's not fast enough to run in real-time. For this reason, the strong classifiers are arranged in a cascade in order of complexity (see Fig. 13.18), where each successive classifier is trained only on those selected samples which pass

Fig. 13.18 Cascade
architecture

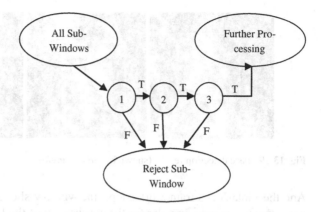

through the preceding classifiers. If at any stage, any sub-window is rejects by a
classifier in the cascade, no further processing is performed and continues to test
next sub-window. Most of the sub-windows have been rejected at the earlier stage
in the cascade, so the evaluation time has been reduced a lots.

This cascade architecture has interesting implications for the performance of the
individual classifiers. Because the activation of each classifier depends entirely on
the behavior of its predecessor, the FPR (false positive rate) for an entire cascade is:

$$F = \prod_{i=1}^{K} f_i. \tag{13.6}$$

Similarly, the detection rate is:

$$D = \prod_{i=1}^{K} d_i. \tag{13.7}$$

To match the FPR typically achieved by detectors, each classifier can get away
with having poor performance. However, at the same time, each classifier in the
cascade needs to be exceptionally capable if it is to achieve suitable detection rates.

Figure 13.19 depicts the face detection results on several 2D intensity images
of ToF depth camera. Since the range data and 2D intensity image are well aligned
on single ToF depth camera, we could easily scratch the corresponding face area in
range data.

13.3.3 Active Shape Model

When a user is ready to use the system to verify his identity, the face detection
process can tell the user is ready or not. Cause if his face is out of the camera or
his pose is not right, there is no need to carry on the following verification process.

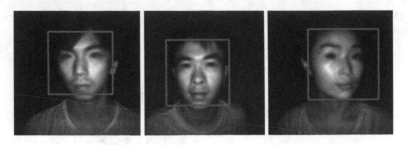

Fig. 13.19 Face detection results (shown by *red rectangle*)

And the Viola-Jones framework adopts the window-sliding strategy, so it may only get a rough face area. In order for the consistency of the feature extraction process, we need to get the exact position and size of the face region before any feature vector can be extracted.

Active shape models (ASMs) are statistical models of the shape of objects which iteratively deform to fit to an example of the object in the new image. The shape of the object is represented by a set of points (which define the shape model). The goal of ASM algorithm is to match the pre-defined model to a new image.

An iterative approach to improve the fit of the instance, X, to an image proceeds as follows:

1. Examine a region of the image around each point X_i to find the best nearby match for the point X_i';
2. Update the parameters (X_t, Y_t, s, θ, b) to best fit the new found points X;
3. Repeat until convergence.

For each point in the shape model, we look along its profiles normal (see Fig. 13.20). If we expect the boundary of the model to correspond to an edge, we can simply locate the strongest edge along the profile. That the new found position gives the new suggested location for the model point.

However, model points are not always corresponded to the strongest edge, they may represent a weaker edge or some other image element. So the best approach is to learn from the training set what to look for in the target image.

In the ASM model, it sample along the profiles normal to the boundaries in the training set, and build statistical models of the grey-level structure. Suppose for a given model point, we sample along the profile k pixels either side of the model point in the i-th training sample. We have $2k + 1$ samples that can be put into a vector g_i. To reduce the effects of global intensity changes we sample the derivative rather than the absolute grey-level values. So the model is much more robust to the global intensity variation. Then we should normalize the sample vector by dividing through the sum of absolute element values. We repeat the same process for each training image, to get a set of normalized sample vector for a given model point. We assume that the distribution of the sample vector as a multivariate Guassian. This gives a statistical model for the grey-level profile about the model

Fig. 13.20 Profile normal

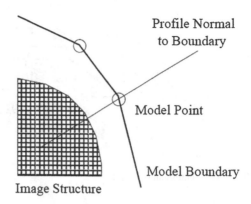

Profile Normal
to Boundary

Model Point

Model Boundary

Image Structure

point. This process is repeated for every point, giving one grey-level model for each point.

During search we sample a profile m pixels either side of the current point $(m > k)$. We then test the quality of fit of the corresponding grey-level mode at each of the $2(m - k) + 1$ possible positions along the sample point and choose the one which gives the best match. We repeated this for each model point, giving a suggested new position for each point. We apply the iteration of the algorithm illustrated before, to update the current pose and shape parameters to best match the model to the new points.

In order to improve the efficiency and robustness of the algorithm, it's implementing in a multi-resolution framework. That means first searching for the object in a coarse image, then refining the location in a series of finer resolution images. This can lead to a faster algorithm. Normally, a Gaussian image pyramid is built (see Fig. 13.21). The base image (Level 0) is the original image. The next image (Level 1) is formed by smoothing the original then sub-sampling to obtain an image with half the number of pixels in each dimension. The next level is formed further smoothing and sub-sampling. Next step, we build statistical models of the grey-levels along profiles normal through each point, at each level of the Gaussian pyramid.

To summaries, the full ASM search algorithm is as follows:

1. Set $L = L_{max}$; L_{max} is coarsest level of Gaussian pyramid to search;
2. While $L >= 0$,
   (a) Compute model point positions in image at level L.
   (b) Search at n_s points on profile either side each current point, n_s is the number sample points either side of current point.
   (c) Update pose and shape parameters to fit model to new points.
   (d) Return (a) unless more than P_{close} of the points are found close to the current position, or N_{max} iterations have been applied at this resolution. P_{close} is proportion of points found within $n_s/2$ of current pos. N_{max} is the maximum number of iterations allows at each level.
   (e) If $L > 0$ then $L = L - 1$.

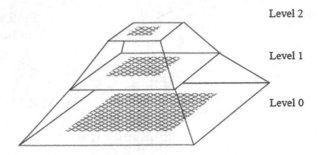

Level 2

Level 1

Level 0

Fig. 13.21 Gaussian pyramid

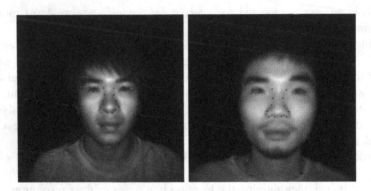

Fig. 13.22 Key points in ToF image

3. Final result is given by the parameters after convergence at level 0.

We train our ToF-ASM model using the 2D amplitude image. And we labeled the feature points by hand (see Fig. 13.22). The feature points correspond to edges and transition part in the facial image. Since the 2D image we used here is quite different with the usual 2D gray image captured by color camera. The pixel value in the amplitude image represents the NIR light reflected back to the camera. So the difference in the boundary between the foreground and background is clearer than the common 2D image. But the details about the eye, nose and mouth region are worse than the common 2D image. The feature points around the face are good enough for face cropping.

We build up a small datasets, which contain five people with several different view of face image. And we label 51 feature points by hand in each face amplitude image. After the ToF-ASM model has been trained using our datasets, we can use this model to crop face region. The result can be found in Fig. 13.23.

For every sample in the data base, we can get the face data. Then we need to extract the feature vector from the face region. We will talk about more this in Chap. 14.

Fig. 13.23 Facial regions
after cropping

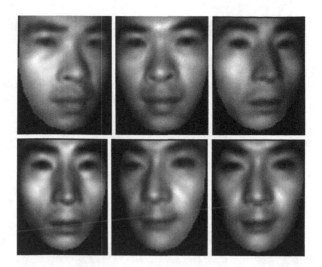

13.4 Experiment and Discussion

We have developed two different systems (Fig. 13.24). The first one—"Capture System", is used to capture data only. You can use this system to find the best parameter to the current scene, test algorithm, recover the 3D facial mesh from a video sequences. The system interface is looks like in Fig. 13.25.

There are four kinds of data from the ToF camera: Distance data, Amplitude data, Intensity data and Mesh data. And the whole interface is refreshed in real-time. You can change the camera status by control buttons. The 3D mesh can be controlled by the mouse, etc., middle button to zoom in/out, left button to rotate.

The control buttons from left to right: button 1 can show the 2D data or hide it; button 2 can play or pause the camera; button 3 is recording button when the camera is playing; using button 4 you can switch recording states between recording a pre-defined number of frames and recording a video without limitation. Press the button 5 you can open the Control panel.

"File" menu (Fig. 13.26) consists of five primary commands:

- "Extract Frame Data", you can extract each frame's data from a large file block, including four types of data.
- "Average Multi-Frame Data", you can average several frames, and get the average data. After this procedure, you can reduce random noise and enhance the data quality.
- "Open 3D model", you can open 3D mesh file, like OBJ file or PLY file.
- "Analyze Video", from a video sequence contains different views of same object that change continuously; you can reconstruct a single 3D Mesh.
- "Save current Frame", when the camera has been opened, if you pause it, you save the current frame to the disk.
- "Exit", quit the program if only when you camera is not connected or paused.

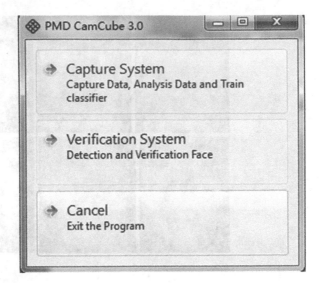

Fig. 13.24 Mode choose

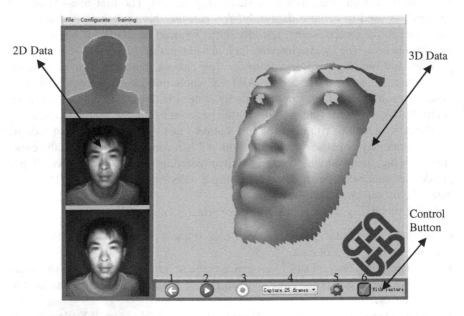

Fig. 13.25 System interface

Just to declare, since our system is using facial image, we can only crop the face region from the whole image. So if the frame data contains no face, there will be an error.

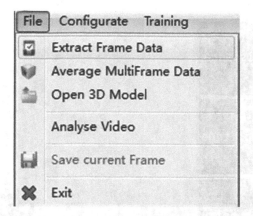

Fig. 13.26 System interface

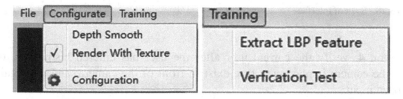

Fig. 13.27 System interface

"Configurate" menu (Fig. 13.27, left) consists of three commands:

- "Depth Smooth", it use the same technique, illustrated in Sect. 13.1, to smooth the mesh data. Strongly recommend, you only use this command when you have crop the specific object from the whole image. Because there may be several unrelated object appear in the same frame data.
- "Render with Texture", you can render the 3D mesh with or without the texture. In this context, texture means the amplitude data for each point.
- "Configuration", open the same control panel like button 5.

"Training" menu (Fig. 13.27, right) consists of some interface for algorithm test. With the "Extract LBP Feature" command, you can extract the LBP feature from the source data, which you can do more analyze and test.

The other one is involved from the first one—"Verification System", focus on our prime purpose: face verification. There are three core part of the second system: register new user's data, update the classifier and face verification based on user's input ID number. The system interface is looks like in Fig. 13.28.

This interface is similar to the capture system. The big difference is the core verification module. Button 1 and Button 2 act exactly the same in the capture system.

Button 3: add new user to the dataset. Only be enabled when the camera is playing;

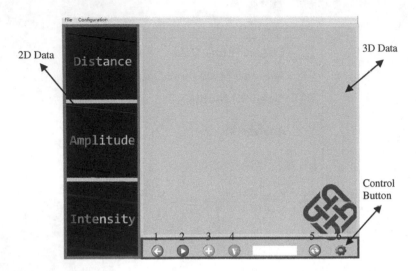

Fig. 13.28 System interface

Button 4: verify the current user after the user has entered his/her ID, only when the camera is playing, a face exist in front of the camera, and the legal ID has been input;

Button 5: input the user ID, double-click it, a dial board will pop up if you don't have a keyboard available;

Button 6: if the user dataset has been updated, you can update the classifier.

"Configuration" menu (see Fig. 13.29) contain two more commands than the same menu in the capture system.

- "Manage Clients": you can manage the information of current registered users, including their name and ID number (see Fig. 13.30).
- "Train New Classifier": after new user has been added, you can update the classifier.

The other important function of the verification system is online registration. When the camera is connected and played, you can press Button 3 to pop up the registration panel (see Fig. 13.31).

After you input the new user's name, you can start to register new user's data, from the mid view, to left view, then right view. You can do this by the instruction in the registration wizard.

After the user is ready, you can press the capture button, the state light will turn red. When the data have been saved to dish, the state light will turn green, and then you can go to the next step (see Fig. 13.32).

For simplicity, the ID system is designed by the registration order. The first registration user's ID is 1. And the last one's ID in our datasets is 43. You can find all these information in the client's information panel.

Fig. 13.29 System interface

Fig. 13.30 Clients info panel

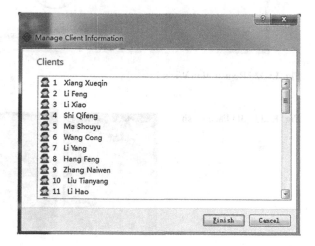

Fig. 13.31 Registration panel

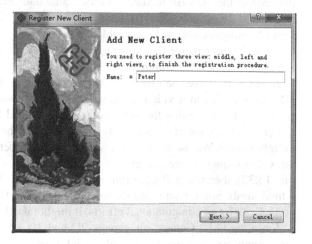

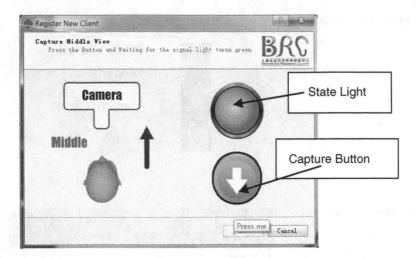

Fig. 13.32 Registration Wizard

Fig. 13.33 3D Face mesh

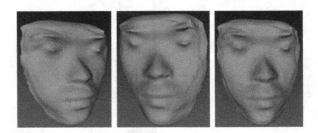

If all new users' data have been successfully saved to the datasets, the system will extract the 3D LBP feature from the frame data and save it automatically. This might take 2–3 s, after that, Button 6 and command "Train New Classifier" under menu "Configuration" will be enabled. Press any of it you can update the classifier. You'd better pause the camera first, to avoid too much pressure to the CPU and memory when the computer is not good enough.

In order to prove the effectiveness of our camera, we can reconstruct a better 3D face mesh from a video sequence of 3D mesh which contains several different views. For recording the video sequence, we need the user to stand before the camera, and face the camera directly at first, turn his/her head left slowly, then turn to right slowly. We use the Poisson Surface Reconstruction algorithm. After we get the video sequence, we can extract several different views of 3D face mesh (see Fig. 13.33), then use ICP algorithm to align all the mesh together, finally, construct a final mesh. Some results are illustrated as follows. But since the 3D reconstruction takes a lot of computational efforts, it might take 3–5 min to finish the process.

Since the PMD Camera can capture 2D and 3D information at the same time, in our system, we fuse these two kinds of information together. The amplitude image

Table 13.1 Comparison with different implementations

	TPR (%)	TNR (%)
3D (2 Layers LBP)	93	95.78
3D (3 Layers LBP)	94.5	96.3
2D & 3D	98.375	97.213

loss so much local and texture information compared to normal 2D gray-scale image. Meanwhile, there are too many noises in the 3D depth data. So using 2D and 3D together, we can get a much more robust system. We also do some experiments if we only use 3D information. You can find the difference of their performance in Table 13.1.

It is obvious that if we combine the 2D & 3D feature together, we can get a more robust and powerful system.

13.5 Summary

In this chapter, we introduced the system principle and architecture of Time-of-Flight (ToF). Based on the principle, we compared the advantages and disadvantages of the ToF capturing systems, and discussed how to use a ToF camera to capture 3D face image. In addition, the source data pre-processing techniques are presented, including denoising, face detection and active shape model. For data denoising, we explained the noise characteristics of ToF data and how to effectively remove the noise. We also presented the face detection method and ASM model.

References

Adelt A, Schaller C, Penne J (2008) Patient positioning using 3d surface registration. In: Proceedings of Russian—Bavarian conference on biomedical engineering, Moscow, pp 202–207, 8–9 July 2008

Bartczak B, Koeser K, Woelk F et al (2007) Extraction of 3D freeform surfaces as visual landmarks for real-time tracking. J Real-Time Image Proc 2(2–3):81–101. doi:10.1007/s11554-007-0042-0

Cai Q, Gallup D, Zhang C et al (2010) 3D deformable face tracking with a commodity depth camera. In: Proceedings of the 11th European conference on computer vision (ECCV'10), Crete, pp 229–242. doi:10.1007/978-3-642-15558-1_17. 5–11 Sept 2010

Cui Y, Schuon S, Chan D et al (2010) 3D shape scanning with a time-of-flight camera. In: IEEE international conference on computer vision and pattern recognition (CVPR'10) 13–18, pp 1–8. doi:10.1109/CVPR.2010.5540082

Fuchs S, Hirzinger G (2006) Extrinsic and depth calibration of TOF-cameras. In: Proceeding of the IEEE conference on computer vision and pattern recognition (CVPR'06), pp 464–468. doi:10.1109/CVPR.2008.4587828

Ganapathi V, Plagemann C, Koller D et al (2010) Real time motion capture using a single time-of-flight camera. In: IEEE international conference on computer vision and pattern recognition (CVPR'10) 13–18, pp 1–8. doi:10.1109/CVPR.2010.5540141

Holte M, Moeslund T, Fihl P (2008) Fusion of range and intensity information for view invariant gesture recognition. In: IEEE international conference on computer vision and pattern recognition workshop on time of flight camera based computer vision (CVPRW'08) 24–26, pp 1–7. doi:10.1109/CVPRW.2008.4563161

Huang X, Ren L, Yang R (2009) Image deblurring for less intrusive iris capture. In: IEEE international conference on computer vision and pattern recognition (CVPR'09) 20–26, pp 1–8. doi:10.1109/CVPR.2009.5206700

Huhle B, Jenke P, Strasser W (2008) On-the-fly scene acquisition with a handy multisensory system. Int J Intell Syst Technol Appl (IJISTA) 5(3/4):255–263 (Issue on dynamic 3D imaging 2008)

Jain HP, Subramanian A (2010) Real-time upper-body human pose estimation using a depth camera, HPL-2010-190

Kim YM, Theobalt C, Diebel J et al (2009) Multi-view image and ToF sensor fusion for dense 3D reconstruction. In: IEEE workshop on 3-D digital imaging and modeling (3DIM'09) 3–4, pp 1542–1549. doi:10.1109/ICCVW.2009.5457430

Koeser K, Bartczak B, Koch R (2007) Robust GPU-assisted camera tracking using free-form surface models. J Real-Time Image Proc 2(2–3):133–147. doi:10.1007/s11554-007-0039-8

May S, Werner B, Surmann H et al (2006) 3D time-of-flight cameras for mobile robotics. In: Proceeding of the IEEE/RSJ international conference on intelligent robots and systems (IROS'06), pp 1578–1586. doi:10.1109/IROS.2006.281670

Penne J, Soutschek S, Fedorowicz L et al (2008) Robust real-time 3D time-of-flight based gesture navigation. In: IEEE international conference on automatic face and gesture recognition 17–19, pp 1–2. doi:10.1109/AFGR.2008.4813326

Rapp H (2007) Experimental and theoretical investigation of correlating ToF-camera systems. Master's thesis, University of Heidelberg, Germany

Soutschek S, Penne J, Hornegger J et al (2008) 3D gesture-based scene navigation in medical imaging applications using time-of-flight cameras. In: IEEE international conference on computer vision and pattern recognition workshop on time of flight camera based computer vision (CVPRW'08) 24–26, pp 1–6. doi:10.1109/CVPRW.2008.4563162

Streckel B, Bartczak B, Koch R et al (2007) Supporting structure from motion with a 3D range-camera. Lect Notes Comput Sci 4522:233–242. doi:10.1007/978-3-540-73040-8_24

Thouis R, Durand F, Desbrun M (2003) Non-iterative feature-preserving mesh smoothing. SIGGRAPH. doi:10.1145/882262.882367

Tong J, Zhang M, Xiang XQ et al (2007) 3D body scanning with hairstyle using one time-of-flight camera. J Comput Anim Virtual Worlds 22(2–3):203–211. doi:10.1002/cav.392

Xiang X, Pan Z, Tong J (2011) Depth camera in computer vision and computer graphics: an overview. J Front Comput Sci Technol 5(6):481–492. doi:10.3778/j.issn.1673-9418.2011.06.001

Zhu J, Yang R et al (2009) Reliability joint depth and alpha matte optimization via fusion of stereo and time-of-flight sensor. In: IEEE international conference on computer vision and pattern recognition 20–26, pp 1–8. doi:10.1109/CVPR.2009.5206520

Chapter 14
3D Face Verification System

Abstract Face recognition systems have the potential to become a key component in various applications like identity authentication, access control, surveillance and security. Compared to conventional biometrics technologies, such as fingerprint and iris imaging, face recognition is non-intrusive and they are easy to acquire. However, face recognition systems do not achieve accuracies of conventional biometrics technologies yet. Furthermore, most existing face recognition systems are computationally expensive and susceptible to inaccuracies caused by variations in illumination, face orientation and partial occlusion of facial features. In this chapter, we will build a 3D face recognition system, develop and investigate suitable methods and techniques for 3D face recognition. We will focus on quality of the 3D data since such data often is very noisy and biased and then crop facial region using window-sliding: strategy and active shape model. After acquiring and pre-processing the 3D data, it is important to extract discriminant and robust features. In this chapter, we will introduce the 3D LBP-AdaBoost (Local Binary Patterns, Adaptive Boosting) method to extract the face features etc. Experiments on real 3D face database will be carried out to test the effectiveness of the method and the 3D face recognition system.

Keywords 3D local binary patterns • Adaptive boosting • Biometrics

14.1 Introduction

The human face has emerged as one of the most promising biometrics. Facial recognition systems have the potential to become a key component in a variety of applications like identity authentication, access control, surveillance and security, or law enforcement (Bowyer 2004; Li 2005). Compared to conventional biometrics technologies, such as fingerprint (Zhao 2010) and iris imaging (Kong 2010), face recognition is non-intrusive and they are easy to acquire. However, face

D. Zhang and G. Lu, *3D Biometrics*, DOI: 10.1007/978-1-4614-7400-5_14,   257
© Springer Science+Business Media New York 2013

recognition systems do not achieve accuracies of conventional biometrics technologies yet. Furthermore, most existing face recognition systems are computationally expensive and susceptible to inaccuracies caused by variations in illumination, face orientation and partial occlusion of facial features.

Due to the availability of the technology, till recently, most research in face recognition has predominantly centered around two dimensional (2D) image processing (Zhao 2003), with a focus on facial feature detection. Although quite decent results has been reported by using 2D face images in constrained environments where illumination is assumed to be constant.

With the development in three-dimensional (3D) information acquisition, 3D systems are popular nowadays. By using of 3D data, some drawbacks in 2D face recognition, such as viewpoint- and illumination-based limitations, would be eliminated (Llonch 2010). Furthermore, 2D face recognition systems are unable to accurately determine the physical dimensions, location and orientation of the face relative to the sensor, whereas 3D sensors typically integrate this information.

The goal of this paper is to develop and investigate suitable methods and techniques for 3D face recognition using range images acquired from a time-of-flight (ToF) 3D depth camera. To this end, we first focus on quality of the 3D data since such data often is very noisy and biased (Frank 2009; Rapp 2008) and then crop facial region using window-sliding strategy (Vioal 2004) and active shape model (Cootes 1995). After acquiring and pre-processing the 3D data, it is important to extract discriminant and robust features. In this paper, we use 3D Local Binary Patterns (3D LBP) (Huang 2006) towards efficient 3D face data representation and these features are then put into Adaboost classifier (Frend 1995; Drucker 1993) for training or testing. Experimental results on real 3D face database show that, by employing pre-processing steps and statistical classification with multiple training and test images of 3D LBP features, 3D face recognition with range images form ToF depth camera provides a feasible and appealing alternative to standard 2D face recognition system.

14.1.1 2D Face Recognition

There are lots of methods and algorithms about 2D face recognition. Since the data collection about 2D face is much more convenient than the 3D face. There are also lots of public 2D database on the websites. Automatic face recognition can be seen as a pattern recognition problem, which is very hard to solve due to its nonlinearity.

Since higher the dimension of the space is, more the computation we need to find a match, a dimensional reduction technique is used to project the problem in a low-dimensionality space. Indeed, the classical Eigenfaces can be considered as one of the first approaches in this sense. For example, an $N \times N$ image I is linearized in a N^2 vector, so that is represents a point in a N^2-dimensional space. However, comparisons are not performed in this space, but a low-dimensional space is found by means of principal component analysis (PCA) technique. The eigenvectors of the covariance matrix is referred to as Eigenfaces (Kirby 1990), represent a base in a low-dimensionality space. When a new image has to be tested, the corresponding Eigenface expansion is computed and compared against the entire database, according to such a distance measure.

The linear discriminant analysis (LDA) (Lu et al. 2003) has been proposed as an alternative to the PCA. It is a supervised learning process, while the PAC is an unsupervised process. Indeed the main aim of the LDA consists in finding a base of vectors providing the best discrimination among the classes, trying to maximize the between-class differences, minimizing the within-class ones. Even if the LDA is often considered to outperform the PCA, an important qualification has to be done. Actually, the LDA provides better classification performances only when there is a wide training set. Another famous method Fisherfaces, the PCA is considered as a preliminary step in order to reduce the dimensionality of the input space, and then the LDA is applied to the resulting space, in order to perform the real classification. In some cases the LDA is applied directly on the input space, as in Chen (2000). Lu (2003) proposed an hybrid between the D-LDA (Direct LDA) and the F-LDA (Fractional LDA), a variant of the LDA, in which weighted functions are used to avoid that output classes, which are too close, can induce misclassification of the input.

The discriminant common vectors (DCV) (Cevikalp 2005) represents a further development of this approach. The main idea of the DCV consists in collecting the similarities among the elements in the same class dropping their dissimilarities. Each class can be represented by a common vector computed from the within scatter matrix. When a new face is tested, the corresponding feature vector is computed and associated to the class with the nearest common vector.

The main disadvantage of the PCA, LDA and Fisherfaces is their linearity. For example, the PCA method extracts a low-dimensional representation of the input data only exploiting the covariance matrix, so that no more than first- and second order statistics are used. Bartlett (2002) shows that first- and second- order statistics hold information only about the amplitude spectrum of an image, discarding the phase-spectrum, while some experiments bring out that the human capability in recognizing objects is mainly driven by the phase-spectrum.

A nonlinear solution to the face recognition problem is given by the neural networks, largely used in many other pattern recognition problems. The advantage of neural classifiers over linear ones is that they can reduce misclassifications among the neighborhood classes. The basic idea is to consider a new with a neuron for every pixel in the image. Also, they need a dimensionality reduction technique before the training process. A first solution to this problem has been given by Cottrell et al. (1990), which introduced a second neural net, which operates in auto-association mode. Lin (1997) presented the Probabilistic Decision Based Neural Network, which they modeled for three different applications. In (Meng 2002), they proposed a hybrid approach, in which, through the PCA, the most discriminating features are extracted and used as the input of a RBF neural network. The RBFs perform well for face recognition problems, as they have a compact topology and learning speed is fast.

The Gabor filter is a powerful tool in image coding, because their capability to capture important visual features, such as spatial localization, spatial frequency and orientation selectivity. And the Gabor filter can be used to extract the main features of human face images, as the formula below shows

$$\psi_{u,v}(z) = \frac{\|k_{u,v}\|^2}{\sigma^2} e^{\left(-\frac{\|k_{u,v}\|^2\|z\|^2}{2\sigma^2}\right)} \left[e^{ik_{u,v}z} - e^{-\sigma^2/2} \right], \quad (14.1)$$

where u and v denote the orientation and scale of Gabor kernels. In (Lades 1993) they proposed a graph-based method. They applied the Gabor Filter to specific areas of the face region which corresponding to a node of a rigid gird. The nodes are linked to form such a Dynamic Link Architecture, so the graph matching strategy can be adopted to calculate the difference between different subjects. Wiskott et al. (1997) further expanded on DLA and developed a Gabor wavelet based elastic bunch graph matching method (EBGM) to label and recognize human faces. Usually, dynamic link architecture is superior to other face techniques, in terms of rotation invariant; however, the matching process is computationally expensive. Perronin et al. (2003) proposed a further deformable model, whose philosophy is similar to the EBGM. They introduced a novel probabilistic deformable model of face mapping, based on bi-dimensional extension of hidden markov model (HMM). A faster wavelet based approach has been proposed by Garcia et al. (2000), which presented a novel method for recognition of frontal views of face under roughly constant illumination.

14.1.2 3D Face Recognition

While there is extensive work on 2D face recognition, 3D face recognition is still a comparatively new research field. But the 2D face recognition methods still encounter some difficulties, like pose, illumination and expression variations. The system performances drop greatly when those difficulties appear. The main advantage of the 3D based approaches is that the 3D model retains all the information about the face geometry. Xu et al. (2004) compared intensity images against depth images with respect to the discriminating power of recognizing people. From their experiments, the authors concluded that depth maps give a more robust face representation, because intensity images are heavily affected by changes in illumination.

One main disadvantage of a face recognition system using range images, however, is the high cost of an industrial high resolution 3D scanner that's often needed to acquire the data. Most of the 3D face recognition works published use such laser or structured-light scanners. Of course the stereographic imaging is a cost-effective way to acquire the range data. But it's well-known that such system needs a robust solution for the correspondence problem and precise calibration.

Two main representations are commonly used to model faces in 3D applications that are 2.5D and 3D images (see Fig. 14.1). A 2.5D image (range image) consists

(a)   (b)

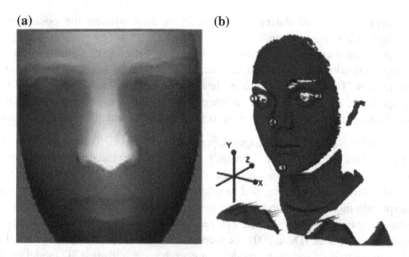

Fig. 14.1 **a** 2.5D Depth image and **b** 3D image

of a two-dimensional representation of a 3D points set (x, y, z), where each pixel in the X–Y plane stores the depth value z. In particular, a 2.5D image taken from a single viewpoint only allows facial surface modeling, instead of the whole head. You can build the 3D head model by taking several scans from different viewpoints during the training stage. The simplest 3D face representation is a 3D polygonal mesh, which consists of a list of points connected by edges. There are two differences between the 3D images and 2.5D images. The one is that 3D images are a global representation of the whole head, while 2.5D images depend on the external appearance as well as environment conditions. Another difference is that the 3D images are not affected by self-occlusions of the face, when the pose is not full-frontal.

For the final task of face recognition for 2.5D images or 3D models, three main methodologies can be identified as:

- Shape matching. This kind of method consists of algorithms that iteratively try to map a 3D point cloud or a 3D mesh directly onto a reference point cloud or reference mesh. The shape matching methods can be seen as pattern recognition methods without feature extraction process. The similarity measure can be optimized by using a sufficiently large number of training samples. But those approaches demand an extensive computational effort and an accurate point-to-point data registration and assume the existence of many correspondences between the reference model and the test data.
- Feature-based methods. The core part of feature-based methods corresponds mostly to that of shape matching. However, different with the first kind of method, not the whole data is processed but appropriate subsets. For example, some important regions, like eye, nose, forehead, or the nose profile of the face could be detected, extracted and processed. Like shape matching methods, the feature-based methods demand a robust image registration, since the

features are selected during a pre-processing step without the possibility to change their value later on.

- Image-based methods attempt to extract the face data subset significant for face recognition with the aid of statistical learning techniques and without any human interaction. There is no or at least less pre-processing steps involved with the image-based methods. So all of the image information is used for statistical analysis. Those methods have been very successful in the context of 2D face recognition.

Approaches based on 2D images supported by some 3D data are identified as 2D-based class methodologies. Generally, the idea to use the 3D faces to improve robustness with respect to appearance variations such as hard pose and facial expression. Blanz (2003) proposed to synthesize various facial variations by using a morphable model that augments the given training set containing only a single frontal 2D image for each subject. The recognition task is achieved measuring the Mahalanobis distance (Duda 2001) between the shape and texture parameters of the models in the gallery and fitting model. Another approach using a 3D model to generate various 2D facial images is given by Lu (2003). They generated a 3D model of the face from a single frontal image. From this 3D model many views are synthesized to simulate new poses, illuminations and expressions. This method achieves a recognition rate of 85 %, outperforming the PCA-based methods on this dataset.

For 3D-based method, one possible approach to gain a correct alignment is by using an acquisition system based on a morphable model, cause it's pre-aligned within a given reference frame, like Ansari's work in Ansari et al. (2003). Starting from one frontal and one profile view image, they use 3D coordinates of a set of facial feature points to deform a morphable model fitting the real facial surface. The recognition task is ten performed calculating the Euclidean distance between some key features points lying on facial surface on mouth, nose and eyes. The iterative closest point (ICP) algorithm (Besl et al. 1992) is often used as an alternative approach alignment models. It could be used to reduce misalignment during the registration phase as well as to approximate the volume difference between two surfaces. Another similar ICP-based algorithm to find a point-to-point correspondence between key features is given by Infanoglu et al. (2004). First, they obtain a dense point-to-point matching by means of a mesh containing points that are present in all faces, so that the alignment is trivially obtained. Then, the Point Set Distance is used to compute the distance between two different clouds of points.

Other kinds of aspect dealing with 3D face recognition concerns the analysis of the 3D facial surface in order to extrapolate information about the shape. Some approaches are based on a curvature-based segmentation detecting a set of fiducial regions. Moreno et al. (2003) used Gaussian curvature to segment the facial regions (see Fig. 14.2).

Bronstein (2003) presented a new method based on a bending invariant canonical representation (see Fig. 14.3), they called canonical image that models deformation resulting from facial expression and pose variations. Besides, Tsalakanidou (2003) proposed an HMM approach to integrate depth data and intensity image.

Adopted the Log-Gabor Filter (Fields 1987) to extract feature on 3D face data. Since the maximum bandwidth of a Gabor filter is limited to approximately one

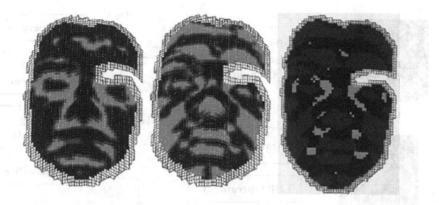

Fig. 14.2 Example of 3D mesh segmentation

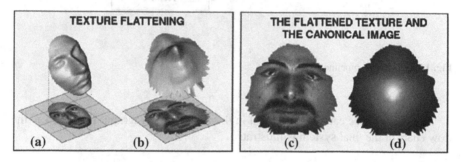

Fig. 14.3 Facial surface representation

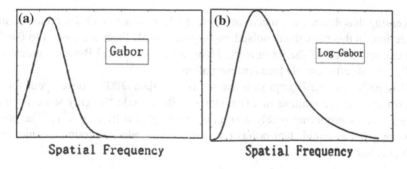

Fig. 14.4 **a** *Gabor* filter response, **b** *Log-Gabor* filter response

octave and Gabor filters are not optimal if one is seeking broad spectral information with maximal spatial localization. Unlike Gabor filter, Log-Gabor filters can be constructed with arbitrary bandwidth and the bandwidth can be optimized to produce a filter with minimal spatial extent. So Log-Gabor filter can cover more bands in the middle and high frequency area (see Fig. 14.4). We can get more robust features to represent the 3D faces.

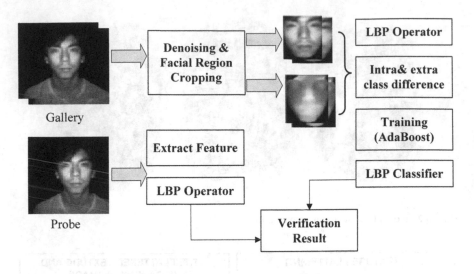

Fig. 14.5 Software pipeline

We will describe each part in detail in the following paragraph. In Fig. 14.5, the flow chart of the final system is illustrated.

14.2 Feature Extraction

Processing has been introduced in Sect. 13.3, now we will introduce feature extraction. In this step, the denoised and segmented 3D faces are passed to the feature extraction part of the framework. Here, we use 3D Local Binary Patterns (3D LBP) towards efficient 3D face representation.

Originally, the LBP approach came from (Ojala 2002), developed for the description and recognition of 2D textures. LBP encode the gray-scale invariant pattern of N neighboring pixels with gray values $X_i, i \in \{0 \ldots N - 1\}$. The neighbors are given as equidistant points on a circle with radius r around a center pixel with gray value c:

$$LBP_N^r = \sum_{i=0}^{N} sig(x_i - c) \cdot 2^i, \qquad (14.2)$$

with $Sig(x) = \begin{cases} 1 & \text{if } x > 0 \\ 0 & \text{otherwise} \end{cases}$

The LBP approach has been successfully applied to 2D face recognition. It has several advantages for face recognition. First, it has high discrimination power by

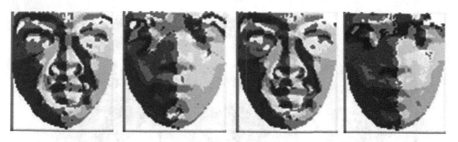

Fig. 14.6 LBP Image of same person at different time

characterizing a very large set of edges. Second, LBP is more robust to lighting variations. Since the noise in the 3D data and pose change in the two sessions, there is more different in the 3D part than 2D part. It can be seen in Fig. 14.6. Third, since histograms are used as features, they are more robust to misalignment and pose variations. But if we apply it directly in the range data, it can only encode the signs of depth differences and thus not adequate for describing 3D faces, since different depth differences on the same point of the facial surfaces distinguishes different faces. For example, if two facial regions of different persons on the same part have the same trends of depth variation, they will generate the same result about binary patterns.

For the computation of 3D LBP values, in our method not only the signs of the gray-scale differences to the neighboring pixels are coded, but also the values themselves are considered. Four bits are used to keep track of a difference value: the lowest bit (level) for sign of the depth difference and 3 bits for the absolute value of the depth difference. Figure 14.7 illustrates some examples of 3D LBP results on the second level.

14.2.1 3D LBP

The original form of LBP operator comes from (Ojala 2002). It labels the pixels of an image by threshold the $3 * 3$ neighborhood of each pixel with the value of the center pixel and considering the results as a binary number. Figure 14.8 shows an example of LBP calculation. And then the 256-bin histogram of the labels computed over a region can be used as a texture descriptor. Each bin can be regarded as a micro-texton. Different types of image structure are codified by different local primitives.

In (Ojala 2002) the LBP operator has been extended to several types with different radius and different sizes. For example, the operator $LBP_{8,1}$ (see Fig. 14.8) uses 8 neighbors on a circle of radius 1. In general, the operator $LBP_{P,R}$ refers to a neighborhood size of P equally spaced pixels on a circle of radius R that form a circularly symmetric set. So there are 2^P different binary patterns that can

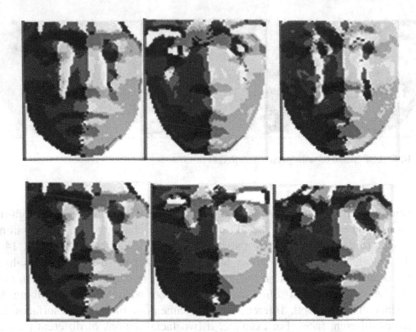

Fig. 14.7 3D LBP results of different person on the second level

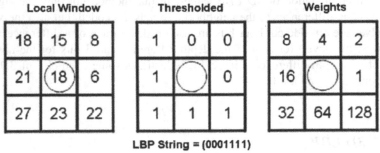

LBP String = (0001111)

LBP Code = 0+0+0+8+16+32+64+128=248

Fig. 14.8 LBP calculation

be formed in the neighbor set. But not all the patterns contain rich information. In (Ojala 2002), they defined the fundamental patterns (they call it uniform patterns) as those with a small number of bitwise transitions from 0 to 1 or 1 to 0. For example, 00000000 and 11111111 contain 0 transitions while 00000110 and 01111000 contain 2 transitions. So the 00000000 and 11111111 are the uniform patterns. Formal definition, we designate patterns that have transition value of at most 2 as "uniform" and propose the following operator for gray-level and rotation invariant texture description. Accumulating the non-uniform patterns into a single bin yields an LBP descriptor with less than 2^P bins. In the computation of the LBP histogram, uniform patterns are used so that the histogram has a separate bin for every uniform pattern and all non-uniform patterns are assigned to s single bin.

$$ALBP_{P,R}^{riu2} = \begin{cases} \sum_{P=0}^{P-1} S\left(g_p - g_c\right) & \text{if} \cup \left(LBP_{P,R} \leq 2\right) \\ P+1 & \text{otherwise,} \end{cases} \quad (14.3)$$

where

$$U(LBP_{P,R}) = |s(g_{P-1} - g_c) - s(g_0 - g_c)| + \sum_{p=1}^{P-1} |s(g_p - g_c) - s(g_{p-1} - g_c)|.$$

$$(14.4)$$

Experimental result shows that uniform patterns account for a bit less than 90 % of all patterns when using the (8, 1) neighborhood and for around 70 % in the (16, 2) neighborhood with texture images. For facial image data sets, it shows that 90.6 % of the patterns in the (8, 1) neighborhood and 85.2 % of the patterns in the (8, 2) neighborhood are uniform in case of pre-processed FERET facial images.

LBP operator has been successfully applied to 2D face recognition. It has several advantages for face recognition. First, it has high discrimination power by characterizing a very large set of edges. Second, LBP is more robust to lighting variations. Third, since histograms are used as features, they are more robust to misalignment and pose variations. But if we apply it directly in the depth data, it can only encoding the signs of depth differences and is not adequate for describing 3D faces, since different depth differences on the same point of the facial surfaces distinguishes different faces. For example, if two facial regions of different persons on the same part have the same trends of depth variation, they will generate the same result about binary patterns.

In order to obtain local correlative features of facial surface, we use the 3D Local Binary Patterns (3D LBP, see Fig. 14.9) to extract feature. Based on the experiment, we can find that 3D LBP not only enhances local properties and details of the texture information of face images, but also can extract local details effectively.

First we apply the 2D LBP to the depth map. According to statistically analysis, more than 90 % of the depth different between points in the local-window are small than 7. And we can know it by instinct, there wouldn't be a big depth change in the small 3 * 3 window. 3 lowest binary units can correspond to the binary number of the absolute value of the depth difference: 0–7 are assigned to higher layer. To summary, the lowest level encodes the sign of the depth difference, and the remaining layers encode the absolute value of the depth difference. In Fig. 14.10, there are some examples about the result of 3D-LBP operator.

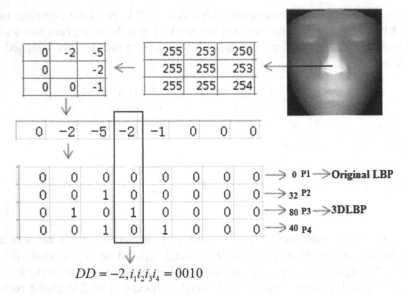

Fig. 14.9   3D-LBP calculation

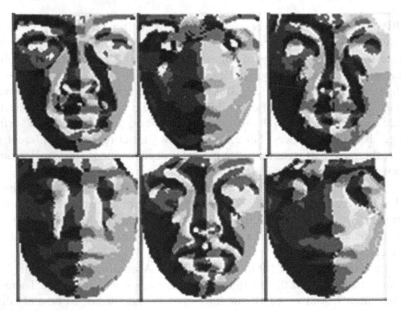

Fig. 14.10   3D-LBP result of different objects

14.2.2 AdaBoost

AdaBoost, short for Adaptive Boosting, is a machine learning algorithm, formu-
lated by Freund and Schapire (Freund 1995; Schapire 1999). It can be used in
conjunction with many other learning algorithms to improve their performance

and learning accuracy. Boosting is a general method which is used to boost the accuracy of any given learning algorithm. Boosting is firstly applied to real-world application for an Optical Character Recognition task, relying on neural networks as base learners in Drucker (1993). AdaBoost often tends not to over fit when running for a large number of iterations. As we mentioned before, in human face detection, AdaBoost was used to train a strong classifier to detect faces in an image.

AdaBoost is an efficient method of producing a highly accurate predication rule by combining a set of rough and moderately inaccurate classifier. In this context, a strong classifier is consisted with a set of weak classifiers. In order to design a strong classifier, there are two issues need to be resolved:

- How is a set of weak classifiers organized into a strong classifier?
- How is a weak classifier selected so that it contributes to a strong classifier?

AdaBoost resolves these two problems and gives the better performance of classification

Pseudo-code for AdaBoost is illustrated as follows:

1. Given the training set $(x_i, y_i), \ldots, (x_n, y_n)$, where x_i is the data of the ith example, and $y_1 \in Y = \{0,1\}$.
2. Initialize the weights $\omega_{1,i} = 1/N$ for each example (x_i, y_i), where N is the total number of training samples.
3. for $t = 1, \ldots, T$ do
4. Train a weak classifier h_t using the weights $\omega_{t,i}$.
5. Calculate the training error with the weak classifier h_t.

6. We define $a_t = \frac{1}{2} In \left(\frac{1-\varepsilon_t}{\varepsilon_t} \right)$;

$$\varepsilon_t = \sum_{i=0}^{N} \omega_{t,i} \, \|h_t(x_i) - y_i\|^2 \tag{14.5}$$

7. And during each iteration, we update the weights like this:
8.

$$\omega_{t,i} = \omega_{t,i} \times \begin{cases} e^{-a_t} & \text{if } h_t(x_i) = y_i \\ e^{a_t} & \text{if } h_t(x_i) \neq y_i \end{cases} \tag{14.6}$$

9. Next, we normalize the weights:

$$\omega_{t+1,i} = \frac{\omega_{t,i}}{\sum_{i=1}^{N} \omega_{t,i}}, \tag{14.7}$$

where the normalization can be expressed by a normalization factor Z_t, so that all $\omega_{t+1,i}$ will keep a probability distribution.

10. end for

After the training process, we can get a final "strong" (also called ensemble) classifier:

$$H(x) = \begin{cases} 1 & \sum_{t=1}^{T} a_t h_t(x) \geq \frac{1}{2} \sum_{t=1}^{T} a_t \\ 0 & otherwise. \end{cases} \tag{14.8}$$

You need to train T "weak" classifiers repeatedly. To train a weak learner, there are many approaches, Perceptron (Rosenblatt 1958), Linear Discriminant Analysis (Martinez 2001), Neural network (Riedmiller 1993), etc. When h_t is trained, if an example (x_i, y_i) is misclassified by the weak classifier h_t, the weight $\omega_{t,i}$ is added in error ε_t. If the example is correct classified, the weight will not be counted in ε_t. The smallest ε_t corresponds to the strongest weak learner h_t. Finally, the strong classifier $H(x)$ is a weighted majority vote of the T weak classifiers where a_t is the weight assigned to h_t. The strong classifier is called ensemble, which refers to a collection of statistical classifiers. Lots of research have shown that an ensemble classifier is often more accurate than any of the single classifier in the ensemble.

Figure 14.11 displays the training process of the "strong" ensemble classifier. There are many points which represent the training data, which form as a random Gaussian distribution $N(0, 1)$. The original training data is not linearly separated, but non-linearly separable.

14.2.3 3D LBP-Boosting

We build our classifier based on LBP operator and AdaBoost. For the purpose of overcome more distortion in the face image, like noise, occlusions and misalignments, we adopt a Robust Local Histogram (RLH) strategy, which takes advantage of the robustness of the accurate local statistical information. Because the LBP description computed over the whole face image encodes only the occurrences of the micro-patterns without any indication about their locations. To overcome this defect, we first divide the facial image into several non-overlapping blocks from which the local binary pattern histograms are computed. Then all the histograms of the local blocks are concatenated as geometric statistics of correlative features for the source data.

AdaBoost procedure essentially learns a two-class classifier; we can solve our verification problem into a two-class one using the intra and extra-class difference. A difference is taken between two 3D LBP histogram feature sets, which is intra-class if the two face images are of the same person, or extra-class if not.

We adopt the AdaBoost algorithm to select the most distinct features. Given the feature vector consisted with LBP histograms, we have two ways to train the weak classifier:

1. Treat the histograms of single block as a sub-feature;
2. Treat a single bin value in the histograms vector as a sub-feature.

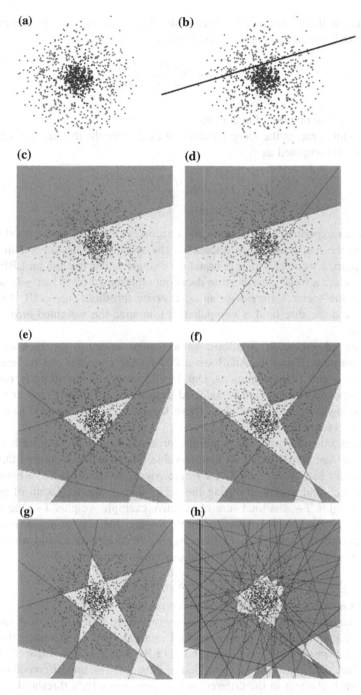

Fig. 14.11 Construction of the ensemble classifier. **a** The original data. **b** The first weak learner.
c $t = 1$. **d** $t = 2$. **e** $t = 5$. **f** $t = 6$. **g** $t = 7$. **h** $t = 40$

We can use the γ^2 distance to measure the distance between 2 histograms, the Chi distance between 2 histograms is defined as follows:

$$CHI(S, M) = \sum_{i=1}^{n} \frac{(S_i - M_i)^2}{S_i + M_i}. \tag{14.9}$$

where n is the bin number of histogram.

At the same time, in the second method, we calculate the distance between two bin features is computed as

$$d(f(x_1), f(x_2)) = \frac{(f(x_1) - f(x_2))^2}{f(x_1) + f(x_2)}, \tag{14.10}$$

Based our experiment, the classification accuracy of the second method is way better than the first one. So we implement the second one in our system. Since in our system, a weak classifier is based on a single scalar feature, an LBP histogram bin value; a weak classification decision, represented by $+1$ or -1, is made comparing the scalar feature with an appropriate threshold (also call "Decision Stump"). And the threshold is computed to minimize the weighted error on the training set.

The key advantage of AdaBoost as a feature selection mechanism is the speed of learning. Using the AdaBoost a 200 feature classifier can be learned in O(MNK) or about 1,011 operations. One key advantage is that in each round the entire dependence on previously selected features is efficiently and compactly encoded using the example weights. These weights can then be used to evaluate given weak classifier in constant time.

The weak classifier selection algorithm proceeds as follows. For each feature, the examples are sorted based on feature value. The AdaBoost optimal threshold for the feature can be computed in s single pass over the sorted list. Four sums are maintained and evaluated during the scan process: the total sum of positive example weights T+, the total sum of negative example weights T−, the sum of positive weights below the current example S+ and the sum of negative weights below the current example S−. And the training error for a threshold which splits the range between the current and previous example in the list is

$$err = \min(S^+ + (T^- - S^-), S^- + (T^+ - S^+)), \tag{14.11}$$

these sums are easily updated as the search proceeds.

In our dataset, for each people, they all register three different view of face (see Fig. 14.12): left, middle and right view. And we train the classifier using this dataset. In the test phase, two sets of selected 3D LBP features are compared, one for the input face data and one for an enrolled face data. Given the difference between two sets, each clement in the different set is compared with a threshold to give a weak decision h_t. In short, we compare the test image with all the example images of same people and take the highest score to make the decision.

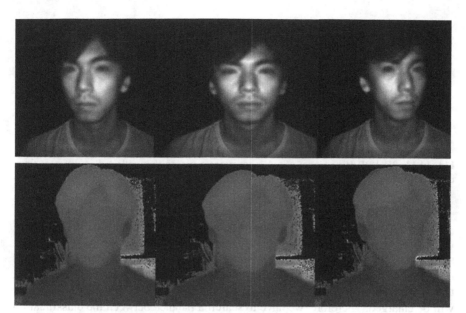

Fig. 14.12 *Three* views of *one* people

14.3 Test Results

Our database contains 270 samples from 45 volunteers with mostly neutral face expression. The 3D face samples were collected for each subject in two separated sessions, and in each session, three samples were collected from the left, middle and right views of 3D face from each subject. The average time interval between the two sessions is two weeks. The PMD CamCube 3.0 ToF depth camera is located in front of the volunteer, while the average distance between camera and volunteer is around 50 cm. To test robustness of our method, for each volunteer, he is not restricted to stare at the centre of camera, but could turn his head to left or right within a context of certain degree (0–10°). The original spatial resolution of the 3D data is 204 × 204. After face cropping, the available 3D face has the spatial resolution about 100 × 100, which is relatively small for traditional face recognition methods.

In feature extraction procedure, we use 3D LBP descriptor to characterize each person.

Figure 14.13 describes three different 3D LBP histograms: two histograms from same person but collected at different time, one histogram for another one. The vertical coordinate in Fig. 14.13 denotes the number of pixels, while the horizontal coordinate corresponds to 'bit-coding' values. For each sample, all local LBP histograms are concatenated together to construct the final histogram, as shown in Fig. 14.13.

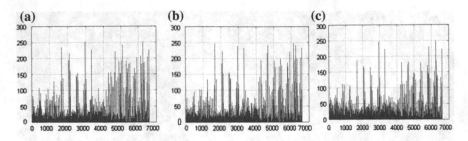

Fig. 14.13 3D LBP Histograms: **a** and **b** from same person, but collected at different time. **c** Another one

Since we adopt the Boosting framework in classification, our final strong classifier is the ensemble of weak classifier. In our method, we use the single node decision tree—"Stump" as the weak classifier, and each weak classifier corresponds to one distinctive feature. It's obvious that, with including more weak classifiers, the strong classifier will be more powerful and robust. However, the training time will be enlarged. Therefore, we have to search a balance between the classification accuracy and the training time consuming.

In RLS strategy, we need to divide the image to several blocks. The block size should be properly chosen. If the block size is too big, the histogram will lose some local information; if it's too small, the block numbers and training time will increase. Before we start the training process, we resize each image to the standard size, which is the average size of all the images in our datasets. In our case, it's just 76 × 88. And there are so many local patches in the image, if we use all of them, we can get the most powerful classifier. But the training time and memory cost will be unaffordable. We need the select the local patch properly. In our experiment, we have tried different size, like 17 × 17, 19 × 19, 21 × 21, and 23 × 23. When the patch size is small, we can get more local information and more patches need to process, when the patch size is big, it loss the local characteristic of facial region even though it's faster to train. Based on our experiment, 21 × 21 can give a powerful classifier and proper training time (less than 120 s).

Figure 14.14 demonstrates how the True Positive Rate (TPR) and True Negative Rate (TNR) vary in verification experiments when the number of weak classifier changes. From Fig. 14.14, it is clearly that when the number of weak classifier is large than 100, the performance will not be improved substantially. Consequently, we choose 100 weak classifiers for classification task in following experiments.

We conducted two types of experiments on the established database: verification and identification. The computational platform is a PC with Windows XP Professional operation system and Core 2 CPU of 1.86 GHz and 2 GB RAM.

The samples collected in the first session are chosen as training samples. In this training phase, the training set of positive examples was derived from

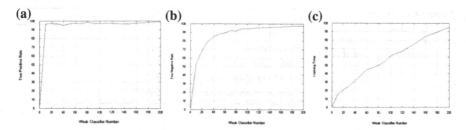

Fig. 14.14   a True Positive Rate. **b** True Negative Rate. **c** Training Time

intrapersonal pairs of 3D LBP histogram features, while the negative set was from extra personal pairs. There were 135 3D faces of 45 people, 3 faces each person, all Chinese. A training set contains about 135 positive and 8,910 negative examples.

A technology evaluation was done with a test set of 135 3D faces, which were collected in the second session. The test set also contained 45 people, with 3 faces per person. None of the test 3D faces were in the training set.

In verification experiments, the class of the input face is known and each of the 3D face was matched with all other 3D faces in the database. If the two 3D faces are from the same class, an intra-class matching or genuine is achieved. On the contrary, an inter-class matching or imposter is found. Using the established 3D face database, there are totally 405 (135 * 3) matches. When using 100 weak classifiers as mentioned before, the TPR (True Positive Rate) and TNR (True Negative Rate) of our method are 97.561 and 94.065 % respectively.

We also conducted the identification experiments on the same 3D face database. In identification, we do not known the class of the face but try to identify which class it belongs to. The identification experiments are designed as follows: we set the samples in training set to be templates and samples in test set as probes. Thus, there are 135 templates and 135 probes respectively. The probes were matched with all the templates, and rank-1 recognition rate was 95.12 % in such experiments.

The computational complexity of our proposed method is relatively cheap. We use Viola-Jones framework for real-time face detection and simple 3D LBP operator for 3D face representation. Since we choose the moderate number of weak classifier to reduce the cost time without affecting the classification accuracy, the run-time of applied AdaBoost classifier is also small. Table 14.1 lists the speed of identification experiment (runtime for each probe evaluated on the same processor as described before), and the matching time of verification experiment is even smaller.

From above experiments, we can conclude that our proposed method is simple but efficient for ToF depth camera based 3D face recognition task. Furthermore, the quick acquisition capacity of ToF depth camera and the fast processing speed of our method make it more suitable for some real-time face recognition applications.

Table 14.1 Average execution times for each processing stage in identification experiment

Processing stage	Execution time (ms)
Denosing	240
Face detection	772
Facial region refinement	452
3D LBP representation	346
AdaBoost classification	289

14.4 Summary

In this chapter, we presented an efficient and robust method for time-of-flight depth camera based 3D face recognition. First, throughout pre-processing steps (i.e., denosing, face detection and facial region refinement), 3D face is cropped from original 3D noisy data. The geometric features of 3D face are then extracted by using 3D LBP approach. Finally, AdaBoost algorithm was explored for classification. We built a 3D face database which contains 270 samples collected from 45 individuals, and conducted a series of verification and identification experiments on it. The experimental results show that our method is well suitable for time-of-flight depth camera based 3D face recognition. Since our 3D system is efficient and robust for face recognition, it could be regarded as a potential alternative to other means of image acquisition for face recognition.

References

Ansari A, Abdel-Mottaleb M (2003) 3D face modeling using two views and a generic face model with application to 3D face recognition. In: Proceedings of IEEE Conference on Advanced video and signal based surveillance (AVSS 2003), pp 37–44. doi: 10.1109/AVSS.2003.1217899

Bartlett MS, Movellan JR, Sejnowski TJ (2002) Face recognition by independent component analysis. IEEE Trans Neural Networks 13(6):1450–1464. doi:10.1109/TNN.2002.804287

Besl PJ, MCKay ND (1992) A method for registration for 3D shapes. IEEE Trans Pattern Anal Machine Intell, 14(2):239–256. doi: 10.1109/34.121791

Blanz V, Vetter T (2003) Face recognition based on fitting a 3D morphable model. IEEE Trans Pattern Anal Mach Intell 25(9):1063–1074. doi:10.1109/TPAMI.2003.1227983

Bowyer K (2004) Face recognition technology: Security versus privacy. IEEE Technol Soc Mag 23(1):9–19. doi: 10.1109/MTAS.2004.1273467

Bronstein A, Bronstein M, Kimmel R (2003) Expression-invariant 3D face recognition. In: Proceedings of audio & video-based biometric person authentication (AVBPA). In lecture notes in computer science 2688:62–69. doi: 10.1007/3-540-44887-X_8

Cevikalp H, Neamtu M, Wilkes M, Barkana A (2005) Discriminative common vectors for face recognition. IEEE Trans Pattern Anal Machine Intell 27(1):4–13. doi:10.1109/TPAMI.2005.9

Chen LF, Liao M, Ko HY, Lin MT, Yu JC (2000) A new lDA-based face recognition system which can solve the small sample size problem. Pattern Recogn 33:1713–1726. doi:10.1016/S0031-3203(99)00139-9

Cootes T, Taylor C, Cooper D, Graham J (1995) Active shape models-their training and application. Comput Vis Image Underst 61(1):38–59. doi:10.1006/cviu.1995.1004

Cottrell GW, Fleming MK (1990) Face recognition using unsupervised feature extraction. In: Proceedings of Intelligence neural network conference, pp 322–325

Drucker H, Schapire R, Simard PY (1993) Boosting performance in neural networks. Int J Pattern Recognit Artif Intell 7(4):705–719

Duda RO, Hart PE, Stork DG (2001) Pattern classification, 2nd edn. Wiley-Interscience, ISBN: 0471056693

Fields D (1987) Relations between the statistics of natural images and the response properties of cortical cells. J Opt Soc Am 4(12):2379–2394. doi:10.1.1.136.1345

Frank M, Plaue M, Rapp H, Köthe U, Jähne B, Hamprecht FA (2009) Theoretical and experimental error analysis of continuous-wave time-of-flight range cameras, Optical Engineering, 48(1):013602–013602-16

Freund Y, Schapire RE (1995) A decision-theoretic generalization of on-line learning and an application to boosting. In Computational Learning Theory: Eurocolt? 5:23–37, Springer. doi: 10.1006/jcss.1997.1504

Garcia C, Zikos G, Tziritas G (2000) Wavelet packet analysis for face recognition. Image Vision Comput 18:289–297. doi:10.1016/S0262-8856(99)00056-6

Huang Y, Wang Y, Tan T (2006) Combining statistics of geometrical and correlative features for 3D face recognition, In: Proceedings british machine vision conference, pp 879–888

Irfanoglu MO, Gokberk B, Akarun L (2004) 3D shape-based face recognition using automatically registered facial surfaces. In: Proceedings of 17th International conference on pattern recognition (ICPR04), Cambridge, pp 183–186. doi: 10.1109/ICPR.2004.1333734

Kirby M, Sirovich L (1990) Application of the Karhunen-Loeve procedure for the characterization of human faces. IEEE Trans Pattern Anal Machine Intell 12(1):103–108. doi:10.1109/34.41390

Kong A, Zhang D, Kamel M (2010) An analysis of iriscode. IEEE Trans on Image Process 19(2):522–532. doi:10.1109/TIP.2009.2033427

Lades M, Vorbruggen JC, Buhmann J, Lange J, Malsburg C, Wurtz RP, Konen W (1993) Distortion invariant object recognition in the dynamic link architecture. IEEE Trans Comput 42:300–311. doi:10.1109/12.210173

Li S, Jain A, Huang T, Xiong Z, Zhang Z (2005) Face recognition applications, In Handbook of Face Recognition 371–390, Springer

Lin S, Kung S, Lin L (1997) Face recognition/detection by probabilistic decision-based neural network. IEEE Trans Neural Networks 8(1):114–132. doi:10.1109/72.554196

Llonch R, Kokiopoulou E, Tosic I, Frossard P (2010) 3D face recognition with sparse spherical representations. Pattern Recogn 43(3):824–834. doi:10.1016/j.patcog.2009.07.005

Lu J, Plataniotis N, Venetsanopoulos N (2003) Face recognition using LDA-based algorithms. IEEE Trans Neural Networks 14(1):195–200. doi:10.1109/TNN.2002.806647

Martinez A, Kak A (2001) PCA versus LDA. IEEE Trans Pattern Anal Mach Intell 23:228–233. doi:10.1109/34.908974

Meng JE, Hock LT (2002) Face recognition with radial basis function (RBF) neural networks. IEEE Trans Neural Networks 13(3):697–710. doi:10.1109/TNN.2002.1000134

Moreno AB, Sanchez A, Velez JF, Diaz FJ (2003) Face recognition using 3D surface-extracted descriptors. In: Proceedings irish machine vision and image, (IMVIP'03), Sep. doi: 10.1.1.96.9489

Ojala T, Pietikainen M, Maenpaa T (2002) Multiresolution gray-scale and rotation invariant texture classification with local binary patterns. IEEE Trans Pattern Anal Mach Intell 24:971–987. doi: 10.1109/TPAMI.2002.1017623

Perronnin F, Dugelay JL (2003) An introduction to biometrics and face recognition. In: Proceedings of IMAGES'03: learning, understanding, information retrieval, medical, Cagliari, Italy, June

Rapp H, Frank M, Hamprecht F, Jähne B (2008) A theoretical and experimental investigation of the systematic errors and statistical uncertainties of time-of-flight cameras. Intl J Intell Syst Technol Appl. 5(3/4):402–413. doi: 10.1504/IJISTA.2008.021303

Riedmiller M, Braun H (1993) A direct adaptive method for faster back-propagation learning:
     The RPROP algorithm. In: Proceeding of international conference on neural network. doi:
     10.1109/ICNN.1993.298623
Rosenblatt F (1958) The perceptron: a probabilistic model for information storage and organiza-
     tion in the brain. Psychol Rev 65:3860408
Schapire RE (1999) A brief introduction to boosting. In: Proceedings of the 16th international
     joint conference on artificial intelligence
Tsalakanidou F, Tzovaras D, Strintzis MG (2003) Use of depth and colour Eigen faces
     for   face   recognition.   Pattern   Recognition   Lett   24(9–10):1427–1435.   doi:10.1016/
     S0167-8655(02)00383-5
Vioal P, Jones M (2004) Robust real-time face detection. Int J Comput Vision 57(2):137–154.
     doi: 10.1023/B:VISI.0000013087.49260.fb
Wiskott L, Fellous JM, Kruger N, Malsburg C (1997) Face recognition by elastic buch graph
     matching. IEEE Trans Pattern Anal Machine Intell 19:775–779. doi:10.1109/34.598235
Xu C, Wang Y, Tan T, Quan L (2004) Depth versus intensity: which is more important for face
     recognition? In: Proceedings of 17th international conference on pattern recognition (ICPR
     2004), vol 4, pp 342–345. doi: 10.1109/ICPR.2004.1334122
Zhao W, Chellappa R, Phillips PJ, Rosenfeld A (2003) Face recognition: a literature Survey.
     ACM Computing Surveys (CSUR) 35(4):399–458. doi:10.1145/954339.954342
Zhao Q, Zhang D, Zhang L, Luo N (2010) High resolution partial fingerprint alignment using
     pore-valley descriptors. Pattern Recogn 43(3):1050–1061. doi:10.1016/j.patcog.2009.08.004

Chapter 15
Book Review and Future Work

Abstract Biometrics recognition, the use of the human physiological and behavioral characteristics for personal authentication, has a long history. In fact, we use it everyday. We commonly recognize people based on their face, voice and gait. Signatures are recognized as an official verification method in legal and commercial transactions. Fingerprints and DNA have been considered effective methods for forensic applications including investigating crimes, identifying bodies and determining parenthood. Recently, more and more effort has been put on developing effective 3D biometrics systems for various security demands.

This book unveils automatic techniques for 3D biometrics authentication, from 3D data imaging, 3D feature extraction and matching, to 3D biometrics system development. It may serve as a handbook of 3D biometrics authentication and be of use to researchers and students who wish to understand, participate, and/or develop a 3D biometrics authentication system. It would also be useful as a reference book for a graduate course on biometrics.

In this chapter, we first recapitulate the contents of this book in Sect. 15.1. Then, Sect. 15.2 discusses the future of 3D biometrics research.

15.1 Book Recapitulation

This book has five parts divided into 15 chapters which discuss the 3D biometrics technology ranging from the hardware design of a 3D biometrics data acquisition, to the algorithm designed for 3D biometrics preprocessing, feature extraction and matching.

Chapter 1 introduces recent developments in biometrics technologies, some key concepts in biometrics, and the importance of developing new biometrics: 3D biometrics.

Chapter 2 gives an overall review of current 3D imaging technologies and their applications in biometrics, including line based and modulated based structured

D. Zhang and G. Lu, *3D Biometrics*, DOI: 10.1007/978-1-4614-7400-5_15,   279
© Springer Science+Business Media New York 2013

light 3D imaging methods, time-of-flight and multi-view 3D data acquisition methods.

Chapter 3 presents a novel method of 3D ear acquisition system by line structured light, and proposes a 3D ear coordinate direction normalization algorithm based on projection density.

Chapter 4 proposes two significant characteristics in 3D ear images: ear-cheek angle and the dissimilarity of the left and right ears of the same person. A modified ICP algorithm is applied for matching the same person's left and right ear. The experimental results show that the ear-cheek angle is unique and stable for each person and distinguishable for different people, which is a useful feature in 3D ear classification and indexing.

Chapter 5 defines five different features in 3D ear, including point, line and area as local feature, and angle and distance as global feature. Then some methods to extract these features are given. The experimental results could illustrate the effectiveness of the features.

Chapter 6 designs and develops a novel three-dimensional palmprint acquisition system based on modulated structured light imaging technology. The acquisition system can obtain 2D and 3D palmprint information at the same time. Also, a 3D palmprint database has been established by using the developed acquisition system.

Chapter 7 explores a 3D palmprint recognition approach by exploiting the 3D structural information of the palm surface. Several types of unique 3D features, including Mean Curvature Image, Gauss Curvature Image and Surface Type, are extracted. A fast feature matching and score level fusion strategy are proposed for palmprint matching and classification.

Chapter 8 proposes three novel global features of 3D palmprints which describe shape information and can be used for coarse matching and indexing to improve the efficiency of palmprint recognition, especially in very large databases. Three proposed shape features are Maximum Depth of palm center, Horizontal Cross-section Area of different levels and Radial Line Length from the centroid to the boundary of 3D palmprint horizontal cross-section of different levels. We treat these features as a column vector and use Orthogonal Linear Discriminant Analysis to reduce their dimensionality. The results demonstrate that the proposed method can greatly reduce penetration rates.

Chapter 9 presents a simple yet efficient scheme for 3D palmprint recognition. After calculating and enhancing the Mean Curvature Image of the 3D palmprint data, we extract both line and orientation features from it. The two types of features are then fused at either score level or feature level for the final 3D palmprint recognition.

Chapter 10 shows a touchless multi-view 3D fingerprint capture system that acquires three different views of fingerprint images at the same time by using cameras placed on three sides. A fingerprint mosaicking method is proposed to splice together the captured images of a finger to form a new image with larger useful print area. 3D finger shape is reconstructed through binocular stereo vision theory. Such 3D information is found very useful in fingerprint alignment. Experimental results show that the proposed mosaic method is more robust to low ridge-valley contrast fingerprint images than available methods.

Chapter 11 studies 3D fingerprint reconstruction technique from touchless multi-view fingerprint images captured, which offers a solution for 3D fingerprint recognition. Several popular used features, such as scale invariant feature transformation (SIFT) feature, ridge feature and minutia, are considered for correspondence establishment.

Chapter 12 explores 3D fingerprint features and their applications for personal identification. We define the 3D finger structural features, such as curve-skeleton, sectional curvatures as Level Zero Fingerprint Features and investigate their distinctiveness for personal identification. A series of experiments is conducted to evaluate 3D fingerprint recognition technique based on our established database with 541 fingers.

Chapter 13 presented an efficient method for 3D face acquiring based on time-of-flight (ToF) depth camera, which is based on measuring the time that light emitted by an illumination unit requires to travel to an object and back to a detector, and is the basis for the development of new range-sensing devices. It is approved that ToF camera is very robust for 3D face imaging.

In Chap. 14, we build a 3D face recognition system, develop and investigate suitable methods and techniques for 3D face recognition. First, throughout pre-processing steps, 3D face is cropped from original 3D noisy data. The geometric features of 3D face are then extracted by using 3D LBP approach. Finally, AdaBoost algorithm was explored for classification. The experimental results show that our method is well suitable for time-of-flight depth camera based 3D face recognition.

15.2 Future Work

We have developed 3D ear (Part II), 3D palmprint (Part III), 3D fingerprint (Part IV), and 3D face (Part V) recognition prototypes already. Each system has its own characteristics, and can handle different real-world challenges accordingly. Although these prototypes are successful, there are still some research aspects that need for further study:

(a) Sensor size
(b) Higher performance
(c) Distinctiveness
(d) Permanence
(e) Privacy concerns

15.2.1 Sensor Size and Cost

Now the sizes of 3D imaging biometrics sensors are larger than corresponding 2D sensors. The same situation lies in their costs. These two factors will greatly affect the application of 3D biometrics technologies. For example, 2D fingerprint sensors are very small now, and it can be used in many fields, such as mobile phone, notebook, electronic lock and so on, but the size of 3D fingerprint sensor based on multi-view is too big to these applications. So there are lots work should be done in the future to control the sensor sizes and costs.

15.2.2 Higher Performance

Higher performance is the main object for all biometrics systems, even though 3D biometrics technologies' performance is better than corresponding 2D technologies (sometimes, these 3D biometrics technologies fuse with 2D technologies), there are still spaces to be improved. More research work can be done on 3D modeling, 3D feature extraction and matching. Another efficient way is to fuse 3D features with others, such as the fusion scheme in Chap. 7 we described the score level fusion and feature level fusion methods. After 2D + 3D feature level fusion we can see that the EER valued by fusing 2D and 3D palmprint are much improved. The EER is decreased from 0.632 to 0.059 % as shown in Table 7.4.

15.2.3 Distinctiveness

It is not uncommon to ask a question about a biometrics, whether it is unique enough for a relatively large user database. For the distinctiveness, here we interest in the information in different 3D biometrics, whether they are sufficiently enough for identifying a person from large population. In other words, can we find out some 3D biometrics candidates from different persons but they are very similar? To investigate the distinctiveness of 3D biometrics, a large scale database is necessary, until now our biggest 3D palmprint database only includes 8,000 samples from 400 palms, and the 3D face database just contains 270 images from 45 volunteers. We should pay more attention to the database construction in the later work, and then build a theoretical model to analyze their distinctiveness.

15.2.4 Permanence

Permanence is another important issue for biometrics identification. Each biometrics has some variations. Even for DNA, mutation is one of the means to change it. Face change depends on our weight, age and living styles, and palmprints have similar situations. But these biometrics characteristics cannot frequently change, some features maybe permanent, and some features are stable in a period of time. Our previous study shows that our method can recognize 3D palmprints collected with a period of time. We need more data to analyze theirs permanence.

15.2.5 *Privacy Concerns*

Privacy concern of a biometrics identifier is another important issue affecting the deployment of a particular biometrics. Like fingerprint technology, some people may be afraid of their fingerprint data being used by third parties for criminal investigation leading them reluctant to adopt it. In order to solve this problem, different aspects should be take care of. In future, we can consider adopting some encryption methods on the 3D biometrics data of each of the communication channels involved so that no one can take the templates out the system for the enrollment on other applications.

Index

D. Zhang and G. Lu, *3D Biometrics*, DOI: 10.1007/978-1-4614-7400-5,
© Springer Science+Business Media New York 2013

Printed in the United States
By Bookmasters